T0257556

Phylogenetics and Phylogeography: Integrated Studies

Phylogenetics and Phylogeography: Integrated Studies

Edited by **Edgar Crombie**

New York

Published by Callisto Reference,
106 Park Avenue, Suite 200,
New York, NY 10016, USA
www.callistoreference.com

Phylogenetics and Phylogeography: Integrated Studies
Edited by Edgar Crombie

© 2015 Callisto Reference

International Standard Book Number: 978-1-63239-511-5 (Hardback)

This book contains information obtained from authentic and highly regarded sources. Copyright for all individual chapters remain with the respective authors as indicated. A wide variety of references are listed. Permission and sources are indicated; for detailed attributions, please refer to the permissions page. Reasonable efforts have been made to publish reliable data and information, but the authors, editors and publisher cannot assume any responsibility for the validity of all materials or the consequences of their use.

The publisher's policy is to use permanent paper from mills that operate a sustainable forestry policy. Furthermore, the publisher ensures that the text paper and cover boards used have met acceptable environmental accreditation standards.

Trademark Notice: Registered trademark of products or corporate names are used only for explanation and identification without intent to infringe.

Printed in the United States of America.

Contents

Preface   VII

Chapter 1   Phylogeography from South-Western Atlantic Ocean:
Challenges for the Southern Hemisphere   1
Graciela García

Chapter 2   Phylogeography of the Mountain Tapir
(*Tapirus pinchaque*) and the Central American
Tapir (*Tapirus bairdii*) and the Origins of the Three
Latin-American Tapirs by Means of mtCyt-B Sequences   21
M. Ruiz-García, C. Vásquez, M. Pinedo-Castro, S. Sandoval,
A. Castellanos, F. Kaston, B. de Thoisy and J. Shostell

Chapter 3   Ecological Factors that Influence Genetic Structure
in *Campylobacter coli* and *Campylobacter jejuni*   55
Helen M. L. Wimalarathna and Samuel K. Sheppard

Chapter 4   The Generation of a Biodiversity Hotspot: Biogeography and
Phylogeography of the Western Indian Ocean Islands   67
Ingi Agnarsson and Matjaž Kuntner

Chapter 5   Hybridisation, Introgression and
Phylogeography of Icelandic Birch   117
Kesara Anamthawat-Jónsson

Permissions

List of Contributors

Preface

Phylogenetic mapping based on geographical measures is perhaps one of the most vital parts of bioscience research. Constant environmental changes have led to a distribution of organisms on land and in water. The imbalance caused by overexploitation of key species not only leading to their extinction, but an imbalance directly affecting the ecosystem. Hence, it becomes the duty of scientists to seek methods to restore balance and rescue the endangered species from non-existence. In order to accomplish this target, species-mapping, history traits and genetic variation must be thoroughly understood. This book helps in the understanding of phylogenetics and phylogeography of a vast range of animals, ranging from disease inducing microorganisms to spiders found in tropical regions, and trees found in the Arctic tundra.

The information contained in this book is the result of intensive hard work done by researchers in this field. All due efforts have been made to make this book serve as a complete guiding source for students and researchers. The topics in this book have been comprehensively explained to help readers understand the growing trends in the field.

I would like to thank the entire group of writers who made sincere efforts in this book and my family who supported me in my efforts of working on this book. I take this opportunity to thank all those who have been a guiding force throughout my life.

Editor

Phylogeography from South-Western Atlantic Ocean: Challenges for the Southern Hemisphere

Graciela García
Evolutionary Genetics Section, Biology Institute,
Faculty of Sciences, UdelaR, Montevideo,
Uruguay

1. Introduction

Since 20 years ago from its emergence in the Evolutionary Genetics area the Phylogeography has experienced explosive growth enhanced by developments in DNA technology, coalescent theory and statistical analysis. Phylogeography is an integrative field of science that uses genetic information to analize the geographic distribution of genealogical lineages, focused within species and/ or between closely related taxa (Avise, 2000). Major impacts have produced at the Biodiversity conservation programs and managements strategies as well as investigating species boundaries and species complex or accessing the patterns and processes of cladogenetic events supporting biological diversity. In a recent review of phylogeography, Beheregaray (2008) identify disparities in research productivity between different regions of the world. He report enormous differences in surface area of habitats, a smaller proportion of studies were conducted on marine organisms than in freshwater organisms. In particular in South America, despite the higher diversity of marine fishes, freshwater fishes were more intensively studied. He proposes that building up of regional comparative phylogeographic syntheses in the Southern Hemisphere is needed to access on the patterns of population history in understudied biotas.

The Southwest Atlantic Ocean region (SWA) generally encompasses the regional waters around Brazil, Uruguay, and Argentina excluding the Falklands/Malvinas Islands (Fig.1). This area, which includes the Patagonian Shelf (Croxall & Wood, 2002) and the convergence of the warm Brazil current from the north and the cold Malvinas current from the south, is characterized by high and consistent levels of primary productivity and supports robust national and international fisheries activities (Campos et al., 1995). Moreover, this region presents unique oceanographic (convergence zone) and physiographic (large continental shelf area) features which also result in high biodiversity of seabird and marine mammal species, as well as sea turtles, all of which use the region for reproduction and/or foraging. In the mid-latitude shelf of eastern South America the discharge of the Plata and the Patos/Mirim lagoons are the major sources of continental runoff. The along-shelf extent of the low salinity plume associated with these systems undergoes large seasonal changes (Piola et al., 2000).The Río de la Plata and its Maritime Front constitute part of the Southwestern Atlantic continental shelf, an ecosystem largely influenced by both Malvinas and Brazilian currents, which conform a confluence in this area. These two water masses have distinct characteristics and constitute a sharp transition zone referred as the subtropical shelf front (Piola et al., 2000).

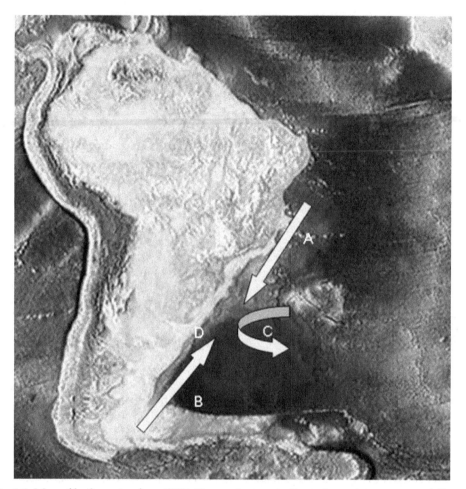

Fig. 1.1. Map of bathymetry from SWA Ocean. Arrows indicate the main currents as follows: A-warm Brazil current from the north, B- the cold Malvinas current from the south, C- Subtropical convergence. D- The subtropical shelf front in the border of the Río de la Plata and its Maritime Front as part of the Southwestern Atlantic continental shelf. (Adapted from http://en.wikipedia.org/wiki/ General_Bathymetric_Chart_of_the_Oceans).

In present work, I present a synthesis of key findings over the scarce phylogeographic research on marine and estuarine organisms mainly fish models from the Southwest Atlantic Ocean region (SWA), to identify research gaps, to drive insight into patterns and processes of taxa diversification associated with the Quaternary Period which have interest not only at regional but also at global marine research. A major goal of present article is to contrast patterns of differentiation among different SWA Ocean endemic fishes and mammals models in which the present distribution in their genetic lineages were strongly nfluenced by Quaternary events. In fact Pleistocene and Post-Pleistocene marine ransgression produced habitat modifications and fragmentations in particular in coastal nd shelf regions from South America (Sprechman, 1978). As a principal consequence of hese events, several rivers and coastal lagoons system were separated from the Atlantic)cean by sand bars, generating associated estuarine environments along South America oast (Montaña & Bossi, 1995). In South America, the marine biogeographic history recorded hree main glaciations events during the Pleistocene (Rabassa et al., 2005). During the last lacial period (LGM), cold marine water moved northwards, changing the Malvinas-Brazil onvergence (38° to 20°).

everal studies have shown that the historical demography of some estuarine fish was rofoundly affected by the Pleistocene glaciations in temperate areas. In fact, during 'leistocene glaciations the sea level had a deep impact on the ecosystems promoting opulation differentiation or bottleneck events. In a more limited timescale (a few centuries) stuarine conditions also appear to be extremely fluctuating, conforming an interface etween marine and riverine influences (Durand et al., 2005). The comparative analysis of hylogeographic patterns would highlighted the relevance of this adapted local narine/estuarine environments to generate and to sustain endemic SWA biodiversity.

!. Assessing the genetic structure in marine populations using molecular lata analysis

pecies are not genetically homogeneous along its distribution, but they are structured into groups of individuals that are typically more or less isolated from each other through or vithout geographic barriers. The resulting pattern for distribution of genetic variation vithin and between populations is referred as the *genetic population structure* of the species Laikre et al., 2005).The basic unit within each of these populations is considered to be a group of individuals characterized by approximate random mating (panmixia) and site enacity to a particular geographic area (Carvalho & Hauser, 1994). This basic unit epresents the "local population", where local refers to the geographic location of the pawning site. As Laikre et al. (2005) explains the genetic population structure of a species vithin a particular geographic area may take a large number of different forms. In fisheries iology the "stock", rather than the local population, is frequently referred as the basic unit or harvest and management. Biologically sustainable management based on knowledge of his structure may reduce the risk for depletion of genetic resources (Laikre et al., 2005). Up o date, most empirical population genetic studies have been based on genetic markers ssumed to primarily reflect selectively neutral DNA variation (most allozymes, nicrosatellites, SNPs and mtDNA markers). Recent molecular developments provide now ncreased opportunities for studying known markers to be located within functional genes of potential importance for fitness (Cano et al., 2008). Among the most used mitochondrial genes and regions in phylogeographic studies the cytochrome b (*cyt-b*) gene has a great utility to resolve phylogenetic relationships in several taxonomic levels (Kocher et al., 1989; Cantatore et al., 1994). This selected molecular marker (*cyt b*) gene and the coalescent in the opulation genetic theory, allowed a retrospective reconstruction of phylogenetic elationships among closely related populations in a past recent time. Moreover, Irwin 2002) argued that mitochondrial DNA are more likely than nuclear markers to show vidence of real barriers to gene flow for two main reasons.: 1) maternally inherited markers ave effective population sizes that are generally one fourth from those of nuclear genes, 2) mitochondrial DNA do not undergo recombination, and hence clear genealogical patterns can be reconstructed.

2.1 Current knowledge in the phylogeography of different endemics marine-estuarine fish ecotypes from SWA Ocean

Estuaries and lagoons are discrete habitats separated by physical and/or ecophysiological barriers that prevent or limit gene flow of estuarine species (Durand et al., 2005). In temperate regions, nearshore marine waters and estuaries act like important nursery areas for several marine teleost species, before they mature and migrate out to deeper waters. In fact, several spawning and nurseries areas for different fish species were detected in rivers, in the Atlantic coastal lagoons system and in the Río de la Plata nearshore environments (García et al., 2008). We have focused present phylogeographic patterns analyses among different selected fish models.

2.1.1 Pelagic fish

In general, pelagic fishes present complex life cycle. According to Sinclair & Iles (1989) in complex life histories, eggs, larvae and juveniles exhibit geographic or spatial distributions that are usually different from those characteristics of the adult phase. Many pelagic fish swim in schools which generally are size-specific, exhibiting a strong preference for similar sized individuals. The school stability can not only have consequences for the population structure and the evolution of the species, but also can have implications for the management purposes of the fishery (Hauser et al., 1998). Despite of the lack of genetic differentiation on a large geographic scale, many studies on pelagic fish detected small but statistically significant genetic differentiation among samples collected from very proximate localities within a short interval of time (Grant, 1985).

Brevoortia aurea. Among migratory pelagic marine fish, the Southwestern Atlantic menhaden *B. aurea* (Clupeidae, Alosinae) represents an important species model to investigate the patterns of genetic differentiation. It is abundant in the Río de la Plata, in the Maritime Front and in the coastal Atlantic lagoons system from southern Brazil, Uruguay and Argentina (García et al., 2008). This taxon is the target of commercial fishery and plays a critical role in the ecology because the individuals are fillter feeders that primarily ingest phytoplankton, providing a direct link between primary productivity and the availability of the forage fish for larger piscivorous predators. In present work, phylogeographic approach based on mitochondrial *cyt*-b gene (Fig. 2.1.1.1) recovered an unexpected high genetic variation in this pelagic marine fish. Bayesian Inference phylogenetic reconstruction using BEAST v.1.5.4 software package (Drummond & Rambaut, 2007) showed a major clade (Fig. 2.1.1.1) including several minor ones with high probability posterior of occurrence.

These minor clades were integrated by structured mixing samples from Rio de la Plata, Atlantic coast and the associated estuarine environments (Mar Chiquita, José Ignacio, Rocha and Castillos coastal lagoons). Present AMOVA analysis revealed a high level of intrapopulation diversity consistent with an extensive and low regional structured population in *B. aurea*. Following Grant & Bowen (1998) for *cyt-b* divergence a rate of 2% per million years can be used to date the nodes in the phylogeny. In present work, *B.aurea* clades from SWA Ocean differentiate since the Plio-Pleistocene, whereas the split from this taxon

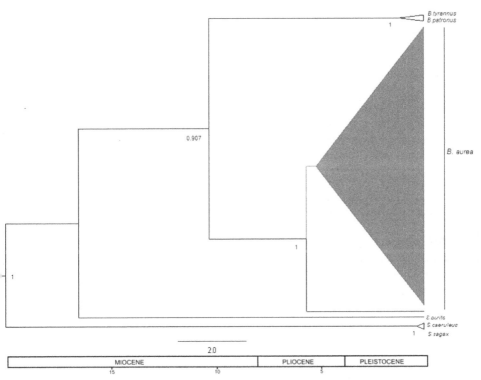

Fig. 2.1.1.1. Bayesian phylogeographic inference framework based on 48 *cyt-b* haplotypes of *Brevoortia aurea* implemented BEAST v.1.5.4 software package (Drummond & Rambaut, 2007). Values below branch nodes refer to highest posterior probability of occurrence for clades. Other Clupeiformes were included as outgoup taxa: *Sardinops sagax, S. caeruleus, Sardinella aurita, Brevoortia patronus* and *B. tyrannus*.The bar below summarize the time-scale divergence dates.

and the northern species of the genus could have occurred during the Miocene. The high values of haplotype and nucleotide diversity (Table 2.1.1.) detected in *B. aurea* were concordant with those reported by Anderson (2007) in four species of the Northern Atlantic menhaden. The present results were also congruent with previous population genetic analyses which suggested the existence of a long-term stability of the *B. aurea* schools, leading to a microgeographical genetic differentiation after the mixture of different stocks in the Río de la Plata and in the Atlantic Ocean from the associated estuarine environments (García et al., 2008). The data revealed that the recruitment of unrelated mtDNA haplotypes carried out by individuals within schools could be occurring in the same nursery areas, revealing the existence of many different maternal lineages. Remarkably, high degree of geographic differentiation was found in the related clupeid *Ethmalosa fimbriata*. In this case, the AMOVA analysis revealed three major geographic units (North, Center and South) for the West African coast populations (Durand et al., 2005). In contrast to the isolation by distance model of population differentiation found in *E. fimbriata* from Western African estuaries (Durand et al. 2005) *B. aurea* shows significant negative values in the Mantel test corroborating a non-association between genetic and geographic distances and excluding the aforementioned possible scenarios (García et al., 2008).

Species	Variable Sites	Parsimony informative Sites	Number of Haplotypes	Haplotype Diversity (SD)	π (SD)	Kimura 2P Distance (Tv+Ts) (SD)	D
B.aurea	209	91	48	1.000 (0.004)	0.037 (0.008)	0.035 (0.008)	-1.883 (P<0.05)
E.anchoita	108	58	10	0.978 (0.003)	0.060 (0.030)	0.086 (0.014)	-1.121 (P>0.10)
L.grossidens	41	18	15	0.838 (0.070)	0.060 (0.013)	0.102 (0.020)	-0.508 (P>0.10)
P. platana	123	20	11	0.908 (0.051)	0.060 (0.037)	0.149 (0.025)	-2.345 (P<0.01)
Ramnogaster sp.	23	4	4	1.000 (0.177)	0.027 (0.006)	0.037 (0.011)	0.373 (P>0.10)

Table 2.1.1. Estimates of population DNA polymorphism in SWA Ocean Clupeiformes. Variables and phylogenetic informative sites among 720 bp in the total data set; Haplotype diversity (h = gene) (Nei, 1987); π = Nucleotide diversity (Nei, 1987); Corrected Kimura 2P distances (Kimura, 1980). D = Neutrality test (Tajima, 1989).

Engraulis anchoita. Fishes of the Engraulidae family, known as anchovies, are widely distributed in tropical and sub-tropical waters. *Engraulis anchoita* is the cold-temperate anchovy in the SWA Ocean. Most anchovies spawn in open coastal areas in the inner continental shelf. Recruitment occurs in protected, shallow areas that offer food and shelter against predators. Adults move during seasons between open coastal areas and bays, where they form large aggregations that are targeted by important fisheries (Araújo et al., 2008). Present phylogeographic analysis based on mitochondrial *cyt-b* sequences in *E. anchoita* from three different collecting regions in the Río de la Plata and the border of continental shelf revealed unexpected large values for haplotype and nucleotide diversity in this taxon. Bayesian Inference phylogenetic reconstruction showed three deepest differentiate clades with high probability posterior of occurrence (Fig. 2.1.1.2). One major clade (Fig. 2.1.1.2b) included most diverse haplotypes belonging to the inner continental shelf (at 35°S-53°W). A sister clade (Fig. 2.1.1.2c) was integrated by haplotypes from the Río de la Plata mouth and outer shelf and the last basal one (Fig. 2.1.1.2a) was represented by haplotypes from inner Shelf (at 36°S-54°W) and Río de la Plata mouth. These results were concordant with the presence of at less three different long-term genetic stocks of *E. anchoita* in the Río de la Plata and its Maritime Front. Therefore the mix of different stocks in the border of the Río de la Plata estuary and SWA Ocean continental shelf at these latitudes may explain the high level of genetic diversity detected in *E. anchoita* according to the present results (Table 2.1.1). Assuming a *cyt-b* rate of 2% of nucleotide divergence these clades could have differentiated since the Plio-Pleistocene. This scenario was consistent with previous data based on *cyt-b* in this taxon (García et al., 2011). Moreover, a haplotype with high probability of occurrence from Rio de la Plata mouth collapse basal in the aforementioned clades representing perhaps a possible ancestral estuarine stock from which populations colonize the Atlantic

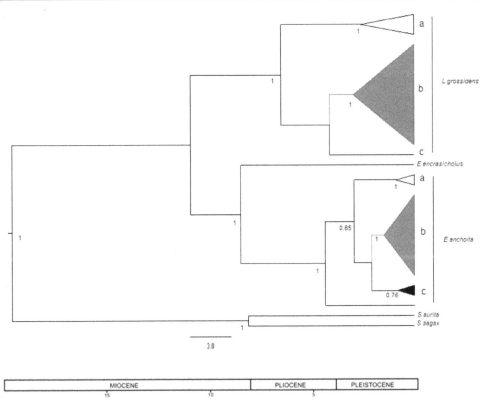

Fig. 2.1.1.2. Bayesian phylogeographic inference framework based on 10 *cyt-b* haplotypes (subclades a,b,c) of *Engraulis anchoita* and 15 *cyt-b* haplotypes (subclades a,b,c) belonging to *Lycengraulis grossidens*. The analysis was implemented in BEAST v.1.5.4 software package (Drummond & Rambaut, 2007). Values below branch nodes refer to highest posterior probability of occurrence for clades. Other Clupeiformes were included as outgoup taxa: *Engraulis encrasicolus, Sardinops sagax, S. caeruleus.* The bar below summarize the time-scale divergence dates.

Ocean shelf environments since the Plio-Pleistocene. In concordance with present data, Late Pleistocene glaciations associated with temperatures, salinity and global shifts in ocean circulation could have had great effect on the historical demography in the Japanese (*Engraulis japonicus*) and Australian (*Engraulis australi*) anchovies, in the northern anchovy *Engraulis mordax* and in the Pacific sardine *Sardinops sagax* (Liu et al., 2006; Lecomte et al., 2004).

Interestingly, present data showed the split of *E. anchoita* and the northern Atlantic-Mediterranean anchovy *Engraulis encrasicolus* since the Miocene, which is consistent with previous hypotheses of differentiation in the genus *Engraulis* (Grant et al., 2005). Although only two species were included in present analysis the genus *Engraulis* appeared as a monophyletic entity whereas *Lycengraulis grossidens* from associated estuarine environments of the SWA Ocean represented a sister clade of *Engraulis*. However, a non overlapping in their respective environments and niches was reported.

Lycengraulis grossidens. The Atlantic sabretooth anchovy occurs in brackish estuaries and adjacent marine areas, penetrating freshwater environments. They form moderate schools usually migrating from estuaries or the sea and spawns in freshwater (Cervigón et al., 1992). The level of population polymorphisms are shown in Table 2.1.1. Present Bayesian Inference based on mitochondrial *cyt-b* sequences in *L. grossidens* from different collecting sites in the Río de la Plata estuary and the associate coastal lagoons between 34°S- 59°W and 36°S-54°W (Fig. 2.1.1.2) revealed the existence of deep differentiate clades. A major clade (Fig. 2.1.1.2b) grouping most haplotypes form all Rio de la Plata and coastal lagoons sites and a single sister haplotype (Fig. 2.1.1.2c) whereas the other minor clade collapse basal (Fig. 2.1.1.2a) to them. The last taxon and minor clade were integrated by samples from a single estuarine site in the border of the Río de la Plata and Atlantic Ocean. Under a *cyt-b* rate of 2% of divergence these clades could have differentiate since the Plio-Pleistocene. Interestingly, the most basal clade could represent perhaps the ancestral estuarine stock from which populations colonize the associate coastal Atlantic lagoons and freshwater environments since the Plio-Pleistocene encompassing major geological and climatic changes in this region.

Platanichthys platana. The Rio Plata sprat is a very small species of fish belonging to the family Clupeidae from SWA Ocean and constitute an endemic monotypic genus. The species inhabit fresh- brackish waters of coastal lagoons, estuaries and the lower reaches of rivers associate to marine areas (Whitehead, 1985). The level of population polymorphisms are shown in Table 2.1.1. Present Bayesian Inference based on mitochondrial *cyt-b* sequences in *P. platana* from different collecting sites in the Río de la Plata estuary and the associate coastal lagoons between 34°S- 59°W and 36°S-54°W (Fig. 2.1.1.3) revealed the existence of two deepest differentiate clades. Assuming a *cyt-b* rate of 2% of divergence, the differentiation of these phylogroups could have occurred since the Plio-Pleistocene. A major clade included all haplotypes from different coastal lagoons (Fig. 2.1.1.3b) and the minor one grouped samples from a single site, Rocha coastal lagoon (Fig. 2.1.1.3a). *Platanychtys platana* clade collapsed in an unresolved basal polytomy with other clupeid genera. These basal split events were dated since the Miocene.

Ramnogaster sp. These small clupeid fish inhabiting littoral areas, estuaries and rivers from Uruguay to Patagonia, constitute endemics taxa from SWA Ocean. They are coastal, pelagic, schooling inshore but not entering freshwaters (Whitehead, 1985). The level of population polymorphisms are shown in Table 2.1.1. Present Bayesian Inference based on mitochondrial *cyt-b* sequences including samples of this genus from different collecting sites in the Río de la Plata estuary and the associate coastal lagoons between 34°S- 59°W and 36°S-54°W (Fig. 2.1.1.3) revealed the existence of two clades with highest posterior probability of occurrence. Both clades (Fig. 2.1.1.3a,b) of *Ramnogaster* sp. could have differentiated since the late Plesitocene. Remarkably, present molecular data does not support the existence of the two described *Ramnogaster* species: *R. arcuata* and *R. melanostoma*. This taxonomic incongruence with previous morphological analysis (Cione et al., 1998) will be clarified in further studies including additional number of samples from both taxa.

Odontesthes perugiae complex. Beheregaray et al. (2002) have investigated patterns of evolutionary divergence and the vicariant history of a group of silverside genus *Odontesthes* distributed in a coastal region of southern Brazil formed during the Pleistocene and Holocene using microsatellite markers and mitochondrial (*D-loop*) DNA sequences. *Odontesthes* is a diverse and widespread genus, with a minimum of 13 species groups

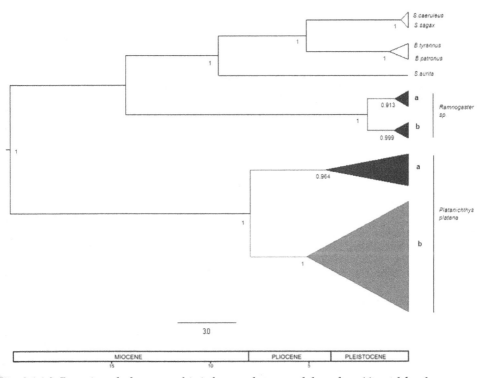

Fig. 2.1.1.3. Bayesian phylogeographic inference framework based on 11 *cyt-b* haplotypes (subclades a,b) of *Platanichthys platana* and 4 *cyt-b* haplotypes (subclades a,b) belonging to *Ramnogaster* sp. The analysis was implemented in BEAST v.1.5.4 software package (Drummond & Rambaut, 2007). Values below branch nodes refer to highest posterior probability of occurrence for clades. Other Clupeiformes were included as outgoup taxa: *Sardinops sagax, S. caeruleus, Sardinella aurita, Brevoortia patronus* and *B. tyrannus*.The bar below summarize the time-scale divergence dates.

distributed in marine, estuarine and freshwater environments of temperate South America (Dyer 1998). The results support the proposal of a silverside radiation chronologically shaped by the sea-level changes of the Pleistocene and Holocene. The radiating lineage comprises a minimum of three allopatric and two sympatric lacustrine species. Four species displayed extremely high levels of genetic variation and some of the most rapid speciation rates reported in fishes. These features were related to a marine- estuarine origin of the radiation. The other interesting feature of *perugiae* is their exceptionally fast rates of speciation. The maximum divergence time based on geology for two of the northern *perugiae* is no more than 5000 years. Based on the model of speciation of silverside fishes of Bamber & Henderson (1988) the aforementioned authors proposed that the rapid divergence of *perugiae* could also be related to its marine- estuarine origin. This model, supported by the reproductive biology and presumed evolutionary patterns of silversides, predicts that physically variable environments, such as estuaries and coastal brackish lagoons, pre-adapt silverside populations to invade, colonize and rapidly speciate into vacant niches in freshwater (Bamber & Henderson, 1988).

2.1.2 Demersal fish

Micropogonias furnieri. The whitemouth croaker *M. furnieri* (Perciformes: Sciaenidae) is a demersal sciaenid with a wide distribution in the central and south-western Atlantic Ocean. Populations of this fish are distributed from the southern Caribbean to the Gulf of San Matías in Argentina. The species occurs along the coast, in the mouth of rivers and lagoons, and also in coastal waters to 50 m (Pereira et al., 2009), being an important commercial fishing resource and at present considered as 'fully exploited'. Recently Pereira et al. (2009), in population genetic analysis based on mitochondrial *D-loop* region detected the presence of two stocks of the *M. furnieri* among three different areas (Bahía Blanca, Río de la Plata and Atlantic Ocean). Remarkably, Río de la Plata locality have the lower values of haplotype and nucleotide diversity as well as demographic signatures of population declination whereas Atlantic Ocean collecting sites showed the highest values of genetic diversity and signatures of a past recent population expansion. More recently, D´Anatro et al. (in press) using seven microsatellite loci, characterized genetic variation of *M. furnieri* in the Río de La Plata estuary and its Maritime Front and in two coastal lagoons. They found three major groups formed by Río de la Plata, Atlantic Ocean and Rocha coastal lagoon populations.

Macrodon ancylodon. The king weakfish *Macrodon ancylodon* (Perciformes: Sciaenidae) is a demersal (bottom-feeding) marine species found in South American Atlantic coastal waters from the Gulf of Paria in Venezuela to Bahia Blanca in Argentina and is economically important because of its abundance and wide consumer species found in South America Atlantic coastal waters from the Gulf of Paria in Venezuela to Bahia Blanca in Argentina (Haimovici *et al.*, 1996). Santos et al. (2006) investigated the phylogeographic patterns in *Macrodon ancylodon* sampled from 12 locations across all its distribution using mitochondrial DNA *cyt-b* sequences together with patterns of morphometric differentiation. Populations of the North Brazil and the Brazil currents, with warmer waters, form a clade (tropical clade) separated by 23 fixed mutations from the populations that inhabit regions of colder waters influenced by the Brazil and Malvinas currents (subtropical clade). No gene flow exists between the tropical and subtropical clades, and most likely also between the two groups of the tropical clade. Distribution of these clades and groups is correlated with flow of currents and their temperatures, and is facilitated by larval retention and low adult migration. Despite the differentiation at the molecular level, fishes analysed from all these current-influenced regions are morphometrically homogeneous. Throughout its range *M. ancylodon* inhabits the same or very similar niche; thus, stabilizing selection probably promotes the retention of highly conserved morphology despite deep genetic divergence at the mitochondrial DNA *cyt-b*.

Cynoscion guatucupa. The striped weakfish (Perciformes: Sciaenidae) widespread pelagic-demersal fish predominantly found on the coast of South America, ranging from Rio de Janeiro, Brazil (22°S), to Chubut province, Argentina (43°S) (Cousseau & Perrotta, 2004). *Cynoscion guatucupa* is considered the second species in commercial importance after the whitemouth croaker *M. furnieri* (Ruarte & Aubone, 2004).

Fernández Iriarte et al. (2011) conducted an analysis based on 365 bp sequence of the mitochondrial control region in four coastal sites located in the SWA Ocean to assess in the pattern of molecular diversity and historical demography of this taxon. Haplotype diversity was high, whereas nucleotide diversity was low and similar at each sample site. AMOVA failed to detect population structure. This lack of differentiation was subsequently observed in the distribution of samples sites in the haplotype network. Fu's Fs was negative and ighly significant while the mismatch analysis yielded a unimodal distribution, indicating a lobal population expansion. This study clearly provides evidence of an older demographic xpansion and no evidence of population phylogeographic structure of *C. guatucupa* from he southwestern Atlantic coast.

.1.3 Endemics benthic and pelagic coastal sharks

quatina guggenheim. Angel sharks (Squatiniformes) are benthic elasmobranchs that inhabit helf and upper slope environments in temperate and tropical regions of the world Compagno, 1984). Among four angel sharks species in the South Atlantic Ocean the angular ngel shark *S. guggenheim* which represent one of the principal coastal fishing resources has a vide geographic distribution from Espíritu Santo (Brazil) to central Patagonia, Argentina Vooren & da Silva, 1991), in waters 10–80 m deep (Cousseau & Figueroa, 2001). Like other enthic elasmobranchs, they tend to have low dispersal capability, which usually result in pecimens from nearby areas which have almost no mixing. As this restricted mixing may roduce different life history parameters, it is important to study possible life history lifferences within angel shark species, even at small geographic scales (Colonello et al., 2007). 'resent phylogeographic approach based on *cyt-b* sequences in *S. guggenheim* from three lifferent collecting sites from the Río de la Plata mouth and its Maritime Front as a part of the iouthwestern Atlantic continental shelf and the Atlantic coast (Fig. 1D) revealed very low values of haplotype diversity (h=0.382, SD=0.078) and nucleotide diversity (π= 0.011, iD=0.003). Tajima test (1989) was significative (D=-2.591, p<0.001) indicating a departure from neutrality in the data set. Despite of the low genetic diversity detected, Bayesian Inference reconstruction grouped the fifteen haplotypes (Fig.2.1.3.1) in two major deepest differentiate :lades. One minor clade integrated by haplotypes from Río de la Plata mouth outer shelf Fig.2.1.3.1a), a second major one formed by samples from Atlantic Ocean coast, Río de la Plata nouth and outer shelf (Fig.2.1.3.1c-d). Remarkably the most frequent and basal haplotype 1 Fig.2.1.3.1b) was shared by samples from the three aforementioned regions. The split between hese deepest clades could have occurred since the Miocene, but subclade differentiation ?vents were dated since Plio-Pleistocene.

Different grouping hypotheses performed in the AMOVA analysis failed to detect iignificant values in the geographic partition of the molecular variance (García et al., 2009). Therefore all present data about genetic stock structure support the hypothesis that S. ;uggenheim constitute a single panmictic unit in the Río de la Plata and its Maritime Front. Low genetic diversity and neutrality departure in angel shark populations would indicate long term population declination since the Pleistocene.

Mustelus schmitti. The narrownose smooth-hound shark, *Mustelus schmitti* is a coastal species ?ndemic to the Southwestern Atlantic Ocean. This species is distributed from Río de Janeiro (Brazil) to Patagonia in Argentina (Chiaramonte & Pettovello, 2000). Pereyra et al. (2010) have used mitochondrial *cyt-b* gene sequences to examine the genetic structure of the narrownose smooth-hound populations within the Río de la Plata and its Maritime Front in the SW Atlantic Ocean. They found no evidence for genetic structure in the analyzed iamples. Low levels of pairwise FST values indicated high connectivity and suggested genetic homogeneity at this geographic range. Additionally, notably low nucleotide and naplotype diversities found in this species could indicate that *M. schmitti* experienced a population bottleneck, recent expansion or selection associated with Pleistocene events. The

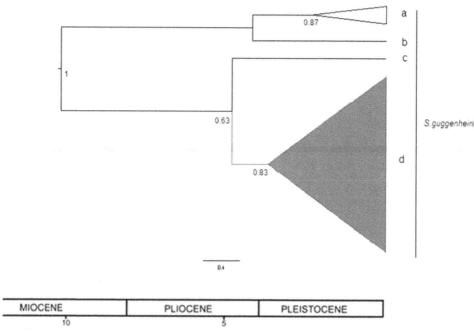

Fig. 2.1.3.1. Bayesian phylogeographic inference framework based on 15 *cyt-b* haplotypes of *Squatina guggenheim* implemented BEAST v.1.5.4 software package (Drummond & Rambaut, 2007). Values below branch nodes refer to highest posterior probability of occurrence for clades. The bar below summarize the time-scale divergence dates assuming a synonymous, per-site divergence rate of 0.0414/million years according to Pereyra et al. (2010).

results presented here indicate that *M. schmitti* exists as a single demographic unit in the Rí de la Plata and its Maritime Front.

2.2 Population genetic structure in marine mammal species models from SWA Ocean

Otaria flavescens. The Southern sea lion distribute in breeding colonies along the entire Atlantic coast. The north-central Patagonian coast is the sea lions most abundant area ir Argentina. Túnez et al. (2010) studied the genetic structure and historical populatior dynamics of the species in five colonies in this area, using the mitochondrial *D-loop* contro region. The distribution of breeding colonies at this smaller geographical scale is also patchy, showing at least three areas with breeding activity. The genealogical relationship between haplotypes revealed a shallow pattern of phylogeographic structure. The analysis of molecular variance showed significant differences between colonies, however, pairwise comparisons only indicate significant differences between a pair of colonies belonging to different breeding areas. The pattern of haplotype differentiation and the mismatch distribution analysis suggest a possible bottleneck that would have occurred 64,000 years ago, followed by a demographic expansion of the three southernmost colonies. Thus, the historical population dynamics of *O. flavescens* in north-central Patagonia appears to be closely related with the dynamics of the Late Pleistocene glaciations.

Pontoporia blainville. The "franciscana" is endemic of South Atlantic coast of South America. The species is under conservation concern because it suffers elevated mortality rates due to incidental captures in fishing nets, and perhaps it is one of the most threatened small cetaceans in this region. Previous morphological and genetic studies have suggested the existence of at least two distinct stocks to the north and south of Santa Catarina Island in Brazil. Secchi et al. (2003) based on mtDNA, morphometrics and population parameters showed that all together provide evidence for splitting the species into four management stocks: two inhabiting coastal waters in Brazil; the third occurring in Rio Grande do Sul State (southern Brazil) and Uruguay, and the fourth inhabiting Argentine coastal waters. However, Lázaro et al. (2004) found fixed differences between a sample from Rio de Janeiro and one from Rio Grande do Sul, in southern Brazil. Using control region of mitochondrial DNA, the authors examined the genetic structure of the species and found no shared haplotypes between Rio de Janeiro and samples from the southern range of the species distribution. However, a phylogenetic analysis suggests that the former population is diphyletic with respect to the southern samples. This suggests that the populations have not been isolated long enough to reach reciprocal monophyly. Furthermore, genetic differentiation is broadly consistent with a simple model of isolation by distance, therefore that appears as an alternative to a model of strict isolation of two stocks. The estimated levels of gene flow are higher among neighboring populations, and decrease when more distant localities are compared. Finally, the molecular data suggest that the "franciscanas" have expanded in Rio de Janeiro.

3. Contrasting phylogeographic patterns in marine-estuarine taxa from SWA Ocean

Like several authors have suggested present analyses reinforce the role of Pleistocene glaciations events to shape the phylogeographic mitochondrial DNA patterns and population structure in marine fish species (Beheregaray et al., 2002; García et al., 2008, 2011). During Pleistocene glaciations the sea level had a deep impact on the ecosystems promoting population differentiation or bottleneck events. In this sense, Fernández-Iriarte et al. (2011) mentioned that the pattern is not necessarily reflected in all coastal fish species, as they could differ in their adaptive response to climate changes (mainly temperature, salinity, marine currents and loss/gain of coastal habitats). Therefore we could expect a different pattern for organism inhabiting pelagic and benthic coastal habitats.

Nevertheless, independently of their adscription to different ecotypes, several major facts not mutually exclusives emerged from present comparative phylogeographic patterns in SWA Ocean among cases analyzed here: 1) population signatures of expansion/declining encompassing environmental changes during the Quaternary, 2) successive colonization-recolonization of marine from estuarine environments in the Quaternary, 3) unexpected and complex population structuring patterns in pelagic fish ecotype, 4) radiation of genetic lineages shaped by the sea-level changes of the Pleistocene and Holocene, 5) Phylogeographic breaks without any specific geographic barriers to gene flow, meanly influenced by Quaternary events and/or of currents and their temperatures , 6) The origin of marine-estuarine endemisms at local and regional level in the SWA Ocean.

1. Genetic footprints of demographic expansion were evident among several studied taxa. In *B. aurea* coastal lagoons have acted like a source of new genetic variants from which

expansion have occurred; *M. furnieri* population stock from Atlantic Ocean showed the highest values of genetic diversity and signatures of a past recent population expansion whereas *C. guatupuca* demographic parameters indicated a global past population expansion. Similar patterns were detected in franciscanas expanding in Rio de Janeiro and in *O. flavescens* showing signature of possible bottleneck, followed by a demographic expansion. It is more difficult to explain the low level of genetic variation and population structure in the benthic angular angel shark *S. guggenheim* and in the narrownose smooth-hound shark, *Mustelus schmitti*. The lowest genetic diversity values detected in these taxa will be indicating the existence of long-term population declination or a past recent population bottleneck associate with major environments and biota changes since the Plio-Pleistocene.

2. Successive colonization-recolonization of marine from estuarine environments promoting dramatic changes in the level of intrapopulation genetic diversity. Several of such environments in the Rio de la Plata and the associate rivers and costal lagoons would acted as glacial refugia from which the recolonization could have occurred. This fact was evident in most Clupeiformes analyzed here. In the other hand, following Avise (1994) the demographic signatures of population declination in *M. furnieri* from the Rio de la Plata stock could be explained if repeated widespread disturbances and founder-flush cycles caused by multiple glacial expansion and retreats, could increase the possibility that the majority of intraspecific mitochondrial diversity originally present would be randomly lost.

3. The unexpected and complex population structuring patterns were evident in pelagic fish ecotypes. In broadly distributed marine fish species that exhibit high vagility, higher dispersal potential during planktonic egg, larval or adult life-history stages low levels of population differentiation are expected. However, in *B. aurea* the analyses suggested the existence of a long-term stability of schools, leading to a microgeographical genetic differentiation after the mixture of different stocks in the Río de la Plata and in the Atlantic Ocean from the associate estuarine environments.

4. Radiation of genetic lineages shaped by sea-level changes of the Pleistocene and Holocene was reported in the pelagic coastal silverside *Odontesthes* (Beheregaray et al., 2002). In this case, estuarine associate environment from coastal lagoon system allowed the divergence of allopatric and lacustrine sympatric populations into different genetic lineages, enhancing cladogenetic events in this genus.

5. Phylogeographic breaks without any specific geographic barriers to gene flow influenced by Quaternary events and/or of currents and their temperatures emerge in different patterns of population genetic structuring analyzed here. Among SWA Ocean taxa, *Platanichthys platana, Ramnogaster* sp., *Lycengraulis grossidens, Engraulis anchoita* as well as in *Squatina guggenheim* deepest population breaks in the phylogeny were evident. As several author have previously mentioned (Lessa et al., 2010) we have detected in these taxa phylogeographic breaks consisting of more than one well-differentiated phylogeographic unit within a region, that differed by more than 3% of sequence divergence from each others. Moreover, *Platanichthys platana, Ramnogaster* sp., *Lycengraulis grossidens* inhabiting the same estuarine environments in coastal lagoons and associate rivers represent a concentration of phylogeographic breaks suggesting a shared history that have occurred at least in part within this region and in the same

scenarios of environment changes during the Quaternary. A remarkably case, is the phylogeographic break in *Macrodon ancylodon* in which the distribution of clades and groups is correlated with flow of currents and their temperatures, and is facilitated by larval retention and low adult migration. Despite of differentiation at the molecular level, fishes analyzed from all these current-influenced regions are morphometrically homogeneous. Nevertheless, only when multiple independent genetic markers show genealogical breaks in the same geographic location it will be possible to conclude that there is a specific geographic cause for the breaks (Avise 2000).

5. The origin of marine-estuarine regional endemisms. In the genera *Brevoortia* and *Engraulis* deepest split in the phylogeny between taxa from the Northern and the Southern Atlantic Ocean were evident since the Miocene. On the other hand, other monotypic and endemic genus *Platanichthys* split from a basal polytomy as phylogenetically separated taxon from the remaining SWA Ocean clupeids since earlier Miocene.

4. General conclusions and future directions

Present comparative analyses reinforce the role of two main factors in the phylogeographic patterns of marine-estuarine fishes and mammals from SWA Ocean: the existence of coastal estuarine associate environments (rivers and lagoon systems) and the ocean currents related with subtropical Convergence, both highly linked to Quaternary dynamics. The shared differentiation patterns detected in present comparative phylogeographic approach would be tested in further analyses including different molecular markers and other taxa from this region. In this sense, more effort is needed to generate not only multi-species, but also multi-locus genetic data-sets to compare and contrast the emerging phylogeographic patterns. Because of the high mutation rate microsatellites will allow researchers to study genetic patterns driven by e.g. fishing pressure and climate change, as well as to obtain more reliable information on gene flow. Migration rates estimated using coalescent-based methods as implemented in different software package often interpreted as reflecting contemporary gene flow may in fact be strongly influenced by historical events, this last emerging from mtDNA phylogeographic patterns. Recent advances in whole and partial sequencing technology (e.g. through 454 pyrosequencing) will allow the development of microsatellite and single nucleotide polymorphism markers more accessible to the world researchers. New and more strong genotype data can be used to identify candidate genes for local ecophysiological adaptations in wild populations reflecting different selection pressure associate to environmental gradients of salinity and temperature. A relatively poor understanding hydrodynamic complex nearshore and offshore patterns in the Rio de la Plata region and in other maritime fronts in the SW Atlantic Ocean represent major challenges to interpret genetic data from taxa that live in shallow waters and disperse within the nearshore environments. The incorporation of both experimental and modeling approaches becomes important for future investigations to interpret the evolutionary response of marine-estuarine species to climatic change.

For most studied species, present analyses support that the metapopulation represents the Management Unit, as well as the inclusion of the associated rivers and coastal lagoons, responsible for the origin of the major genetic diversity, in high priority conservation programs in further Marine Protected Area Systems, which must be integrated at regional level in SWA Ocean. In the other hand, the existence of a long-term population declination or a past recent population bottleneck, alert about the degree of vulnerability to overexploitation in endemics SWA Ocean sharks, as well as, in other important species of Sciaenidae in commercial fisheries. The emerging results in population genetic structure in taxa from the SWA Ocean, Río de la Plata and its Maritime Front will allow to optimize and to develop technical guidelines for the best fishery management stocks as different fishing units in commercial and artisanal fisheries in the region. This knowledge may be increase in next years for most of the fish resources along SWA Ocean area. The molecular genetic markers constitute robust tools, allowing the management of species and stocks and to supervise the long term fishery sustainability.

5. Acknowledgments

The author grateful G. García grants furnished by projects PDT_DINACYT 07/12 and PDT_DINACYT 71/01 for molecular population research in marine and estuarine fishes. The author is also grateful to R. Scorza for suggestions on the manuscript. G. García research is also supported by SNI researcher program (ANII-Uruguay).

6. References

Anderson, J.D. (2007). Systematics of the North American menhadens: molecular evolutionary reconstructions in the genus *Brevoortia* (Clupeiformes: Clupeidae). *Fishery Bulletin*, Vol.205, No.3, (July 2007), pp. 368–378, ISSN 0090-0656

Araújo, F.G.; Silva, M.A.; Santos, J.N.S. & Vasconcellos, R.M. (2008). Habitat selection by anchovies (Clupeiformes: Engraulidae) in a tropical bay at Southeastern Brazil. *Neotropical Ichthyology*, Vol.6, No.4, pp. 583-590,ISSN 1679-6225 Avise, J. C. (1994). *Molecular Markers, Natural History and Evolution*. Chapman and Hall, New York.

Avise, J.C. (2000). *Phylogeography: The History and Formation of Species*. Harvard University Press, Cambridge, Massachusetts

Bamber, R.N. & Henderson, P. (1988). Pre-adaptive plasticity in atherinids and the estuarine seat of teleost evolution. *Journal of Fish Biology*, Vol. 33, Supplement A (December 1988), pp.17–23

Beheregaray, L. (2008). Twenty years of phylogeography: the state of the field and the challenges for the Southern Hemisphere. *Molecular Ecology*, Vol.17, No. 17, (September 2008), pp. 3754-3774, ISSN:09621083

Beheregaray, L.B.; Sunnucks, P. & Briscoe, D.A. (2002). A rapid fish radiation associated with the last sea level changes in southern Brazil: the silverside *Odontesthes perugiae* complex. *Proceedings of the Royal Society of London. Series B, Biological Sciences*, Vol. 269, No. 1486, (January 2002), pp. 65–73

Campos, E.J.D.; Miller, J.L.; Müller, T.J. & Peterson, R.G. (1995). Physical oceoanography of the Southwest Atlantic Ocean. *Oceanography*, Vol. 8, No. 3, pp. 87-91

Cano, J.M.; Shikano, T.; Kuparinen, A. & Merilä, J. (2008). Genetic differentiation, effective population size and gene flow in marine fishes: implications for stock management. *Journal of Integrated Field Sciences*, Vol.5, (March 2008), pp. 1-10 antatore, P.; Roberti, M.; Pesole, G; Ludovico, A; Milella, F; Gadaleta, M.N. & Saccone, C. (1994). Evolutionary analysis of cytochrome b sequences in some Perciformes: evidence for a slower rate of evolution than in mammals. Journal of Molecular Evolution, Vol.39, pp. 589-597, ISSN: 0022-2844

arvalho, G.R. & Hauser, L. (1994). Molecular genetics and the stock concept in fisheries. Reviews in Fish Biology and Fisheries, Vol. 4, No. 3, pp. 326-350, ISSN: 0960-3166

ervigón, F.; Cipriani, R.; Fischer, W.; Garibaldi, L.; Hendrickx, M.; Lemus, A.J.; Márquez, R.; Poutiers, J.M.; Robaina, G. & Rodriguez, B. (1992). Fichas FAO de identificación de especies para los fines de la pesca. Guía de campo de las especies comerciales marinas y de aguas salobres de la costa septentrional de Sur América. FAO, Rome. pp.513

hiaramonte, G.E. & Pettovello, A.D. (2000). The biology of Mustelus schmitti in southern Patagonia, Argentina. Journal of Fish Biology, Vol. 57, No.4, pp. 930–942

ione, A.L.; Azpelicueta, M.M. & Casciotta, J.R. (1998). Revision of the clupeid genera Ramnogaster, Platanichthys and Austroclupea (Teleostei: Clupeiformes). Ichthyological Exploration Freshwater, Vol. 8, No. 4, pp. 335-348

olonello,J.H.; Lucifora, L.O. & Massa, A. (2007). Reproduction of the angular angel shark (Squatina guggenheim): geographic differences, reproductive cycle, and sexual dimorphism. ICES Journal of Marine Sciences, Vol.64, No.1, pp.131–140

ompagno, L.J.V. (1984). Sharks of the world. An annotated and illustrated catalogue of shark species known to date. Part 1. FAO species catalogue 4, FAO Fisheries Synopsis 125, pp.1–249

ousseau, M.B. & Figueroa, D.E. (2001) Las especies del género Squatina en aguas de Argentina (Pisces: Elasmobranchii: Squatinidae). Neotrópica, Vol. 47, pp. 85–86

ousseau, M.B. & Perrotta, R.G. (2004). Peces marinos de Argentina: biologia, distribucion, pesca. INIDEP, Mar del Plata. pp. 163, ISBN 987-20245-4-5

roxall, J.P, & Wood, A.G. (2002) The importance of the Patagonian Shelf for top predator species breeding at South Georgia. Aquatic Conservation: Marine and freshwater ecosystems.Vol. 12, No.1, pp. 101-118

)'Anatro, A.; Pereira, A. & Lessa, E.P. Genetic structure of the white croaker, Micropogonias furnieri Desmarest 1823 (Perciformes: Sciaenidae) along Uruguayan coasts: contrasting marine,estuarine, and lacustrine populations. Environmental Biology of Fishes (In press).

)rummond, A. & Rambaut, A (2007). BEAST: Bayesian evolutionary analysis by sampling trees. BMC Evololutionary Biology, Vol. 7, (November 2007), pp. 214

)urand, J.D.; Tine, M.; Panfili, J.; Thiaw, O.T. & Laë, R. (2005). Impact of glaciations and geographic distance on the genetic structure of a tropical estuarine fish Ethmalosa fimbriata (Clupeidae, S. Bowdich, 1825). Molecular Phylogenetic and Evolution, Vol. 36, (August 2005), pp. 277–287, ISSN:1055-7903

)yer, B. (1998). Phylogenetic systematics and historical biogeography of the neotropical silverside family Atherinopsidae (Teleostei: Atheriniformes). In: L.R. Malabarba; R.E. Reis; R.P. Vari; Z.M. Lucena & C.A.S .Lucena (Ed.) Phylogeny and classification of Neotropical fishes. Part 6.Atherinomorpha, pp. 519-536. Edipucrs, Porto Alegre, Brasil

ernández Iriarte, J.P;Alonso, M.P; Sabadin, D.E.; Arauz, P.A. & Iudica, C.M. (2011). Phylogeography of weakfish Cynoscion guatucupa (Perciformes: Sciaenidae) from the southwestern Atlantic. *Scientia Marina*, Vol. 75, No.4, (December 2011), pp 701-706, ISSN: 0214-8358

García, G.; Vergara, J. & Gutiérrez, V. (2008). Phylogeography of the Southern Atlantic menhaden *Brevoortia aurea* inferred from mitochondrial cytochrome b gene. Marine Biology, Vol. 155, No.3, (September 2008), pp. 325–336, ISSN 0025-3162

García, G.; Pereyra, S.; Oviedo, S; Miller, P. & Domingo, A. (2009) Estructura genética del angelito (*Squatina* spp.) en el Río de la Plata y su Frente Marítimo. Extended Abstract. Proccedings VII Simposio de Recursos Genéticos para América Latina y el Caribe. INIA Carrillanca, Pucón Chile, pp 173-174, ISBN/ISSN: 978-956-7016-35-8

García, G., Martínez,G., Retta, S., Gutiérrez, V.,Vergara, J. & Azpelicueta, M.M. (2011) Multidisciplinary identification of clupeiform fishes from the Southwestern Atlantic Ocean.International. *Journal of Fisheries and Aquaculture*, Vol. 2, No.4 (March 2011), pp. 41-52, ISSN 2006-9839

Grande, L. (1985). Recent and fossil clupeomorph fishes with materials for revision of the subgroups of clupeoids. *American Museum of Natural History*, Vol. 181, pp. 235–372

Grant, W.A.S. & Bowen, B.W. (1998). Shallow population histories in deep evolutionary lineages of marine fishes: insights from sardines and anchovies and lessons for conservation. *Journal of Heredity*, Vol. 89, No.5, pp. 415–426, ISSN 0022-1503

Grant, W. S.; Leslie R. W. & Bowen, B. W. (2005) Molecular genetics assessment of bipolarity in the anchovy genus *Engraulis*. *Journal of Fish Biology*, Vol. 67, No. 5, (November 2005), pp. 1242–1265

Haimovici, M.; Martins, A.S. & Vieira, P.C. (1996) Distribuição e Abundância de peixes teleósteos demersais sobre a plataforma continental do sul do Brasil. *Brazilian Journal of Biology*, Vol. 5, No.1, pp.:27-50. ISSN 0034-7108

Hauser, L.; Carvalho, G.R. & Pitcher, T.J. (1998) Genetic structure in the Lake Tanganyika sardine *Limnothrissa miodon*. *Journal of Fish Biology*, Vol. 53, supplement A, pp. 413–429

Irwin, D.E.(2002). Phylogeographic breaks without geographic barriers to gene flow. *Evolution*, Vol. 56, No.12, pp. 2383–2394

Kimura, M. (1980). A simple method for estimating evolutionary rate of base substitutions through comparative studies of nucleotide sequences. *Journal of Molecular Evoution*, Vol.16, pp. 111–120

Kocher, T.D.; Thomas, W. K.; Meyer, A.; Edwards, S. V.; Paabo, S.; Villablanca, F. X. & Wilson, A. C. (1989).

Dynamics of Mitochondrial DNA Evolution in Animals: Amplification and Sequencing with conserved Primers. *Proceeding National Academy Sciences*.USA, Vol. 86, (August 1989), pp. 6196-6200

Laikre,L.; Palm, S. & Ryman, N. (2005). Genetic Population Structure of Fishes: Implications for Coastal Zone Management. *Ambio: A Journal of the Human Environment*, Vol. 34, No.2, pp. 111-119, ISSN: 0044-7447

Lazaro, M.; Lessa, E.P. & Hamilton, H. (2004). Geographic genetic structure in the franciscana dolphin (*Pontoporia blainvillei*). *Marine Mammal Science*, Vol.20, No. 2, pp. 201-204

Lecomte, F.; Grant, W.S.; Dodson, J.J.; Rodriguez-Sanchez, R. & Bowen, B.W. (2004). Living with uncertainty: genetic imprints of climate shifts in East Pacific anchovy

(*Engraulis mordax*) and sardine (*Sardinops sagax*). *Molecular Ecology*, Vol.13, (August 2004), pp. 2169-2182

Lessa, E.P.; D'Elia, G. & Pardiñas, U. (2010). Genetic footprints of late Quaternary climate change inthe diversity of Patagonian-Fueguian rodents. *Molecular Ecology*, Vol.19, No.15, (July 2010) pp. 3031-3037

Liu, J.X.; Gao, T.X.; Zhuang, Z.M.; Jin, X.S.; Yokogawa, K. & Zhang, Y.P. (2006). Late Pleistocene divergence and subsequent population expansion of two closely related fish species, Japanese anchovy (*Engraulis japonicus*) and Australian anchovy (*Engraulis australis*). *Molecular Phylogenetics and. Evolution*, Vol.40, No.3, (May 2006), pp. 712-23

Montaña, J. R .& Bossi, J. (1995). *Geomorfología de los humedales de la cuenca de la Laguna Merín en el departamento de Rocha*. UDELAR. Facultad de Agronomía, Montevideo, No. 2, pp. 1-32

Nei, M. (1987). *Molecular Evolutionary Genetics*. Columbia University Press, New York. pp. 505.

Pereira, A.N.; Márquez, A.; Marin, M. & Marin, Y. (2009). Genetic evidence of two stocks of the whitemouth croaker *Micropogonias furnieri* in the Río de la Plata and oceanic front in Uruguay. *Journal of Fish Biology*, Vol. 75, No.2 (September 2009), pp. 321-331

Pereyra, S.; García, G.; Miller, P.; Oviedo, S. & Domingo, A.. (2010) Low genetic diversity and population structure of the narrownose shark (*Mustelus schmitti*). *Fisheries Research*, Vol.106, No.3, pp. 468-473, ISSN: 01657836

Piola, A. R; Campos, E. J. D.; Möller Jr., O. O.; Charo, M. & Martinez, C. M. (2000). Subtropical shelf front off eastern South America. *Journal of Geophysical Research*, Vol. 105, pp. 6566- 6578, ISSN 0148-0227

Rabassa, J.; Coronato, A.M. & Salemme, M. (2005). Chronology of the Late Cenozoic Patagonian glaciations and their correlation with biostratigraphic units of the Pampean region (Argentina). *Journal of South American Earth Sciences*, Vol.20 No.1-2, pp. 81-103

Ruarte, C. & Aubone, A. (2004). La pescadilla de red (*Cynoscion guatucupa*), análisis de su explotación y sugerencias de manejo para el año 2004. *INIDEP, Vol.* 54, pp. 1-15

Santos, S.; Hrbek, T; Farias I.P.; Schneider, H & Sampaio, I. (2006). Population genetic structuring of the king weakfish *Macrodon ancylodon* (Sciaenidae), in Atlantic coastal waters of South America: deep genetic divergence without morphological change. *Molecular Ecology*, Vol.15, pp. 4361-4373.

Secchi, E.R.; Danilewicz, D & Ott, P.H. (2003). Applying the phylogeographic concept to identify franciscana dolphin stocks: implications to meet management objectives. *Journal of Cetacean Research and Management*, Vol. 5, pp. 61-68

Sinclair, M. & Iles, T.D. (1989). Population regulation and speciation in the oceans. *Journal du Conseil International pour l'Exploration de la Mer*, Vol. 45, pp. 165-175

Sprechman, P. (1978). The paleoecology and paleogeography of the Uruguayan coastal area during the Neogene and Quaternary. *Zitteliana* Vol.4, pp.3-72

Tajima, F. (1989). Statistical method for testing the neutral mutation hypothesis by DNA polymorphism. *Genetics*, Vol.123, No. 3, (November 1989), pp. 585-595

Túnez, J.I.; Cappozzo, H.L.; Nardelli, M.& Cassini, M.H.(2010). Population genetic structure and historical population dynamics of the South American sea lion, *Otaria flavescens*, in north-central Patagonia. *Genetica*, Vol. 138, No. 8, pp. 831-41

Vooren, C.M. & da Silva, K.G. (1991). On the taxonomy of the angel sharks from southern Brazil, with the description of *Squatina occulta* sp. n. Brazilian Journal of Biology, Vol.51, No.3, pp. 589-602

Whitehead, P.J.P. (1985). FAO species catalogue. Clupeoid Fishes of the world (suborder Clupeoidei). An annotated and illustrated catalogue of the herrings, sardines, pilchards, sprats, anchovies and wolfherrings. Part 1 Chirocentridae, Clupeidae and Pristigasteridae. FAO Fisheries Synopsis 125, FAO, Rome, pp. 1-303.

2

Phylogeography of the Mountain Tapir (*Tapirus pinchaque*) and the Central American Tapir (*Tapirus bairdii*) and the Origins of the Three Latin-American Tapirs by Means of mtCyt-B Sequences

M. Ruiz-García* et al.

*Molecular Genetics Population- Evolutionary Biology Laboratory,
Genetics Unit, Biology Department, Science Faculty,
Pontificia Javeriana University, Bogota DC,
Colombia*

1. Introduction

The Perissodactyla order is a very old group of mammals (around 60 Millions years ago, MYA). In the fossil record, there are representative specimens from five main superfamilies (Tapiroidea, Rhinocerotoidea, Chalicotheroidea, Equoidea and Brontotheroidea) including 14 different families (Savage and Long 1986; Holbrook 1999), although the phylogenetic relationships among these superfamilies are not well resolved. The order had its maximum diversity peak during the Eocene, but during the upper Oligocene, 10 of the 14 families became extinct (Radinsky 1969; Froehlich 1999; MacFadden 1992; Metais et al. 2006). Currently, only three superfamilies and three families are present (Tapiridae, Rhinocerotidae and Equidae).

The first species of the Tapiroidea superfamily appeared in the last phase of the Paleocene and in the lower Eocene (55 MYA) at the same time as the original species of Equidae and Chalicotheriidae and before the apparition of Brontotheriidae and Rhinocerotidae, which appeared close to the end of the Eocene. Some of these original Tapiroidea forms (*Heptodon*, Helaletidae family from the lower Eocene of Wyoming, Radinsky 1963, 1965), showed very similar morphologic resemblances to the current tapirs (*Tapirus*). Some families in the fossil record, such as Hyrachyidae (*Hyrachyus*) of the middle Eocene (45 MYA) of North America

C. Vásquez[1], M. Pinedo-Castro[1], S. Sandoval[2], A. Castellanos[3], F. Kaston[4], B. de Thoisy[5] and J. Shostell[6]
*Molecular Genetics Population- Evolutionary Biology Laboratory, Genetics Unit, Biology Department,
Science Faculty, Pontificia Javeriana University, Bogota DC Columbia*
[4]Tapir Preservation Fund. Bogotá DC, Colombia
[2]Fundación Espíritu del Bosque. c/ Barcelona 311 y Tolosa, Quito, Equador
[4]Fundación Nativa, Colombia
[3]Association Kwata, BP 672, 97335 Cayenne cedex, French Guiana
[6]Biology Department, Penn State University-Fayette, Uniontown, Pennsylvania, USA

and Europe have been conformed to a transition family between Tapiroidea and Rhinocerotoidea (Radinsky 1967, 1968). However, other authors (Holbrook 1999; Colbert 2005) consider Tapiroidea as a taxon which excludes forms related to Rhinocerotoidea. Several families within Tapiroidea are well recognized by all the authors. These are the cases of Deperetellidae (Middle Eocene-Lower Oligocene from Asia; Tsubamoto et al. 2005), Lophiodontidae (Lower and Upper Eocene from Europe and Middle Asian Eocene; McKenna and Bell 1997), Lophialetidae (Middle-Upper Asian Eocene; McKenna and Bell 1997), and Helaletidae (Lower Eocene-Middle Oligocene from North America and Upper Eocene-Lower Oligocene from Asia).

The current Tapiridae family (Gray 1821) is composed of a unique genus, *Tapirus* (Brünnich 1772). The oldest fossil records of this family are dated from the Oligocene of Europe (33-37 MYA), and their fossils have been frequently found in Europe, North America and Asia (Hulbert 1995). Following Radinsky (1965), the Tapiridae descended from the Helaletidae, through the genus *Colodon*. Colbert (2005) defined the Tapiridae family as the clade conformed by the most recent common ancestor of *Protapirus* until the current *Tapirus*. The family has been around since the Lower Oligocene and includes the genera *Protapirus* and *Tapirus*, *Miotapirus* (North-America), *Megatapirus* (Asia), *Tapiravus* (North-America), *Tapiriscus* (Europe), *Eotapirus* (Europe), *Palaeotapirus* (Europe) and *Plesiotapirus* (Asia).

The oldest record of *Tapirus* comes from the European Oligocene, where the fossil remains are found until the Pleistocene (McKenna and Bell, 1997). In North America, the *Tapirus* records indicate that they were present in the Middle Miocene through the present (Hulbert 1995), while for Asia the records indicate that *Tapirus* has been in existence since the Lower Miocene (Deng 2006). Around 20 different *Tapirus* species are recognized for North-America, Europa and Asia.

Of the current four species, two are present in South-America (*T. terrestris* and *T. pinchaque*), one in Central America (*T. bairdii*) and another is present in Asia (*T. indicus*). *T. terrestris* is widely distributed across a great part of South America, including Colombia, Venezuela, Surinam, Guyana, French Guyana, Ecuador, Peru, Bolivia, Brazil, Paraguay and Argentine. *T. pinchaque* is geographically found in the Northern and Central Andes, and is adapted for living at high altitude mountains, in Venezuela, Colombia, Ecuador and northern Peru. *T. bairdii* is distributed from south-eastern Mexico and throughout all of Central America (excluding El Salvador) to the western Andes, in the Colombian and Ecuadorian Chocó. *T. indicus* lives on a very fragmented area of Vietnam, Cambodia, Burma, Sumatra, Thailand, Malaysia and Toba islands (Brooks et al. 1997).

From a molecular genetics point of view, only a few works have been published with *Tapirus*. The first one was the work of Ashley et al. (1996), where the genetics relationships among the *Tapirus* species were analyzed by means of the mitochondrial Cytochrome Oxidase subunit II (mtCOII) gene sequences collected from *T. terrestris*, (two samples) *T. bairdii* (two samples) *T. indicus* (two samples) and *T. pinchaque* (one sample). The second work, completed by Norman and Ashley (2000), included a new sequence of *T. pinchaque*. The authors partially sequenced the 12S rRNA for the eight *Tapirus* specimens that they studied. These two mt genes that were used by previous studies contradict each other so that there is no consensus of the relationship between South/Central American tapirs.

Jorton and Ashley (2004a,b) published two DNA microsatellite works with wild and aptive *T. bairdii* populations. Very recently, a new work on the genetics biogeography of *T. errestris* has been published (Thoisy et al. 2010) showing dispersion of this species from the western Amazon to the rest of South America.

However, no study has included a large amount of samples of the three Latin American *apirus* species. Among the molecular markers relevant for phylogeography, biosystematics, nd genetic structure studies in mammal populations, the mtCyt-b gene is commonly used (Patton et al. 2000; Cortez-Ortiz et al. 2003). Herein, we sequenced 201 individuals belonging o the three *Tapirus* species presented in South America (*T. terrestris, T. pinchaque*) and in central America (*T. bairdii*) for the mtCyt-b.

The main aims of the present study were as follows: 1- To determine the gene diversity evels for the three Latin America *Tapirus* species at the mtCyt-b gene and the degree of genetic divergence among these three species; 2- To analyze the possible demographic historical changes (population expansions or bottlenecks) in the three Latin American *apirus* species; 3- To provide new data on the phylogenetics relationships of the three *apirus* species; 4- To search for possible correlations among the time splits among the haplotypes found within *T. pinchaque* and within *T. bairdii* (for *T. terrestris*, this is shown elsewhere) and 5- To analyze the possible spatial genetic structure in two species (*T. pinchaque* and *T. bairdii*).

2. Material and methods

A total of 201 *Tapirus* samples were analyzed. Of these samples, 141 belonged to *T. terrestris* from different regions of Colombia [41 animals; one from Bajo Sinú-Tierra Alta, (Córdoba Department), 2 from Mesay River (Caquetá Department), one from Fondo Canaima, Vichada Department), 18 from Leticia to San Juan de Atacuarí (Amazonas Department), 7 from Eastern Colombian Llanos (Meta Department), 3 from Pto. Inirida (Guania Department), 3 from Palomino River-Sierra Nevada de Santa Marta and 6 from Antioquia Department)], Venezuela (5 from El Zulia, Maracaibo), French Guiana (11 from Carnopi River), Ecuador (7 animals; 4 from Limoncocha, Sucumbios and 3 from Coca, Sucumbios), Peru [30 animals; one from Arica (Curaray River), 2 from Napo River (Nueva Vida and Mazán), 7 from Nanay River, one from Requena (Ucayali River), one from Bretaña (Canal del Puhinauva-Ucayali River), 4 from Pucallpa (Ucayali River), and 15 from Pto. Maldonado Madre de Dios River)], Bolivia (11 animals; 9 from Mamoré River, one from Chimoré River and one from Villa Bella at the Beni River), Brazil [24 animals; 2 from Yavarí River, 12 from Tabatinga, 2 from Negro River, 2 from Santarem (Pará state) and 6 from Amazon mounth Pará state)], Paraguay (one animal from Hernandarias) and Argentina (4 animals from the Yungas area in Salta-Jujuy) (Figure 1). Additionally, one animal from the Barcelona Zoo Spain), 5 animals from the Cincinnati Zoo (Ohio, USA) and one animal of unknown origin were also analyzed. More details of origins of these samples are recorded in Ruiz-Garcia et al. (2012b). Of the remaining samples, 30 belonged to *T. pinchaque* and 30 to *T. bairdii*. The geographical origins of the *T. pinchaque* samples in Columbia were as follows: six samples from Los Nevados National Park at the Risaralda Department, 14 samples from the Tolima Department (one from the Resguardo Vereda La Bella-Planadas, Marquetalia; three from

Gaitania; four from the Vereda San Miguel-Planadas; two from the Ereje-Blanco River basin, one from La Azulena; one from Peñas Blancas; and two from the Saldaña River basin), and two samples from the Huila Department (Vereda Marengo). Samples within Ecuador came from Papallacta (two samples, Provincia Napo), Cosagua (one sample, Provincia Napo), La Bonita (one sample, Provincia Sucumbios) and Sangay National Park (four samples, Las Culebrillas locality). The samples of this very elusive and rare species were composed of small pieces of skins, bones and teeth of specimens collected by F. Kaston and M. Ruiz García in Colombia and A. Castellanos in Ecuador.

The 30 *T. bairdii* samples were collected as follows: 14 animals were sampled at the Darien region in Panamá (near to the Columbian frontier), nine animals were sampled at the Braulio Carrillo National Park in Costa Rica, three exemplars were sampled at the Petén region in Guatemala and another four specimens were sampled near Campeche in the Mexican Yucatán. These samples consisted of small amounts of blood (the Darien samples) and hairs with roots (for the remainder animals). Figure 2 shows the sampling localities for *T. pinchaque* and *T. bairdii*.

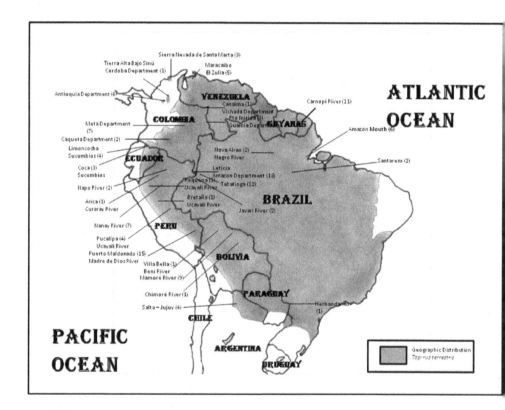

Fig. 1. Map of the geographical distribution of *Tapirus terrestris* and the sampling points. Into parentheses, the number of individuals sampled in each point.

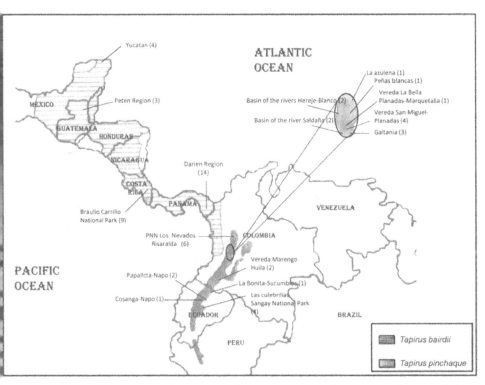

Fig. 2. Map of the geographical distribution of *Tapirus bairdii* and *Tapirus pinchaque* and the sampling points. Into parentheses, the number of individuals sampled in each point.

2.1 Molecular analyses

DNA from teeth, bones, muscle, and skins, were obtained with the phenol-chloroform procedure (Sambrook et al. 1989), while DNA samples from hair and blood were obtained with 10% Chelex® 100 resin (Walsh et al. 1991). Amplifications for mtCyt-b gene were performed with the primers L7 (5' ACC AAT GAC ATG AAA AAT CAT CGT T 3') - H6 (5' TCT CCA TTT CTG GTT TAC AAG AC 3'), which had been designed for perissodactyles (Tougard et al. 2001). The PCRs were performed in a 50-μl volume with reaction mixtures including 10 μl of 10x Buffer, 7 μl of 25mM MgCl₂, 2 μl of dNTPs (dNTP Mix Promega, 40mM), 4 μl (100 μM) of each primer, one unit of Taq DNA polymerase (Gotaq, Promega), 2 μl of DNA from blood, skin or muscle tissue or 2-10 μl of DNA from hairs and teeth and a variable quantity of H₂0. PCR reactions were carried out in a Geneamp PCR system 9600 (Perkin Elmer) and in an iCyclerTM BioRad thermocycler. The temperatures employed were as follows: 94 °C for 5 minutes, 35 cycles of 50 s at 94 °C, 50 s at 53 °C and 1.5 minutes at 72 °C and a final extension time for 10 minutes at 72 °C. All amplifications, including positive and negative controls, were checked in 2 % agarose gels, employing the molecular weight marker φX174 DNA digested with *Hind* III and *Hinf* I and HyperLadder IV and the gels were visualized in a Hoefer UV transiluminator. Those samples that amplified were purified using membrane-binding spin columns (Qiagen). The double-stranded DNA was directly sequenced in a 377A (ABI) automated DNA sequencer. The samples were sequenced in both directions and all the samples were repeated to ensure sequence accuracy.

2.2 Data analyses

2.2.1 Genetic diversity and heterogeneity analyses

The statistics employed to determine the genetic diversity among the three neotropical tapir species were the number of polymorphic sites (S), the number of haplotypes (H), the haplotypic diversity (H_d), the nucleotide diversity (π), the average number of nucleotide differences (k) and the θ statistic by sequence.

Different tests were carried out to measure genetic heterogeneity, and possible gene flow estimates, among these tapir species. These tests were those of Hudson et al. (1992a,b) (H_{ST}, K_{ST}, K_{ST}^*, Z, Z*), Hudson (2000)'s Snn test and the chi-square test on the haplotypic frequencies with permutation tests with 10,000 replicates as well as the G_{ST} statistic from the haplotypic frequencies and the γ_{ST}, N_{ST} and F_{ST} (Hudson et al. 1992a) statistics from the nucleotide sequences.

2.2.2 Demographic genetics analyses in the three Neotropical tapir species

Diverse strategies were used to determine possible demographic changes across the natural history of the three Neotropical tapir species. The procedures employed were as follows: 1- The mismatch distribution (pairwise sequence differences) was obtained following the method of Rogers and Harpending (1992) and Rogers et al. (1996). Two theoretical curves were obtained (population growth and bottleneck show characteristic signatures in histograms yielding the relative frequencies of individual pairs that differ by i nucleotide sites), one assuming a constant population size and another assuming a population expansion with a θ_0 (= $2N_{e0}\mu$). N_{e0} is the female effective number before growth and μ is the mutation rate per generation for the population before the expansion. And, θ_1 (= $2N_{e1}\mu$) with N_{e1} as the female effective number after growth for the same population after the expansion and when $\tau = 2\mu t$ (t = number of generations). This is the time elapsed from the population expansion in a mutational temporal scale. The empirically observed distribution was compared to these two theoretical curves. Some coefficients were used to determine the similarity between the observed and the theoretical curves. These were the raggedness rg statistic (Harpending et al. 1993; Harpending 1994), the Mean Absolute Error (MAE) between the observed and the theoretical mismatch distribution (Rogers et al. 1996) and the R_2 statistic of Ramos-Onsins and Rozas (2002). 2- The Fu & Li D and F tests (Fu and Li 1993), the Fu F_S statistic (Fu 1997) and the Tajima D test (Tajima 1989a), originally created to detect natural selection affecting DNA sequences, were also used to determine possible population size changes (Simonsen et al. 1995; Ramos-Onsins and Rozas, 2002). All these statistics, tests, and analyses were obtained by means of the DNAsp 4.1.03 and Arlequin 3.1 programs.

2.2.3 Phylogenetic analyses and molecular temporal splits among the *Tapirus* species

The mtCyt-b sequence alignments were carried out manually and with the DNA Alignment program (Fluxus Technology Ltd).

o reconstruct the phylogeny and the split times of the three neotropical tapirs, several nalyses were undertaken. The first was the application of the FindModel program to etermine, among 28 different evolutionary nucleotide models, which one was the most robable for the tapir sequence set.

A Bayesian procedure was employed with the BEAST v. 1.4.8 program (Drummond and ambaut 2007) to determine the phylogenetic relationships and the temporal splits among he three neotropical tapirs. In this analysis, we employed the sequences of 11 *T. bairdii*, 14 equences of *T. pinchaque*, 40 sequences of *T. terrestris* and one sequence of *T. indicus*. We educed the number of exemplars in this analysis to obtain some manageable trees. These nimals basically represented all the haplotypes found. The precise and detailed genetic elationships among the 141 *T. terrestris* can be observed in Ruiz-García et al. (2012b). We erformed this analysis to estimate the time to most recent common ancestor (TMRCA) for he different tapir clades found. Analysis was performed using a GTR (General Time Reversible) model of nucleotide substitution with gamma distributed rate variation among ites, and four rate categories (GTR+G) because it was determined to be the better model using the FindModel program. We employed diverse calibration points following Ashley et l. (1996) and Norman and Ashley (2000) and several paleontological records (Patterson and Pascual 1968; Simpson 1980; Webb 2006) in the same Bayesian tree. For this, we used a ombination of a temporal split between the ancestor of *T. indicus* and the ancestor of the hree neotropical tapirs of 18 ± 1 MYA (mtCOII gene; Ashley et al. 1996) and a temporal eparation of 2.7-3.1 ± 0.5 MYA between the ancestors of *T. terrestris* and *T. pinchaque*, during he Great American Biotic Interchange after the reestablishment of the land bridge between North and South America. A Yule process of speciation and a relaxed molecular clock with n uncorrelated log-normal rate of distribution was assumed (Drummond et al. 2006). Results from the two independent runs (20,000,000 generations with the first 2,000,000 discarded as burn-in and parameter values sampled every 100 generations) were combined. The Bayesian trees showed the posterior probability values, which provide an assessment of he degree of support of each node on the tree, the estimations of temporal splits in each node and the lower and upper 95 % highest posterior densities (HPD). Monophyletic onstraints were imposed for nodes that were used to calibrate the evolutionary rates. To nalyze the autocorrelation tree (ACT) and effective sample size for parameter estimates ESS), the program Tracer version 1.4 (Rambaut and Drummond 2007) was employed. The inal tree was estimated in the TreeAnnotator v1.4.5 software and visualized in the FigTree . 1.2.2 program.

To estimate other possible independent divergence times among the haplotypes found in he four *Tapirus* species, a Median Joining Network (MJ) (Bandelt et al. 1999) was applied by means of the software Network 4.2.0.1 (Fluxus Technology Ltd). Once the haplotype network was constructed, the ρ statistic (Morral et al. 1994) was estimated. This statistic measures the age of an ancestral node in mutational units. This value is transformed into years by multiplication with the mutation rate. Additionally, the standard deviation (σ) was alculated (Saillard et al. 2000). The ρ statistic is unbiased and highly independent of past demographic events. These events could have influenced the shape of a given evolutionary ree, but these events only influence the error of the time estimated and do not increase or decrease the time. We employed a mutation rate of 5.6 x 10^{-3} substitutions per site per million years. It was a mean value estimated by Ruiz-García (unpublished result) for a variety of neotropical species (pink river dolphin, Andean bear, deer, and several genera of Neotropical Primates as *Saimiri, Cebus, Alouatta, Ateles, Lagothrix* and *Aotus*). For *T. pinchaque*, this mutation rate represented one mutation each 196,881 years and for *T. bairdii* and the set of all the *Tapirus* sequences taken together, this equaled one mutation each 203,384 years.

2.2.4 Spatial genetics analyses applied to *Tapirus pinchaque* and to *Tapirus bairdii*

Several strategies were applied to determine if *T. pinchaque* and *T. bairdii* presented some significant spatial genetic trend because this could help to understand the evolutionary events that have determined the natural history of these two species. These strategies were as follows:

1. A Mantel's test (Mantel 1967) was used to detect possible overall relationships between a genetic matrix among individuals (Log-Det genetic distance with different pattern heterogeneity among lineages and different rates among sites with a Gamma distribution, Nei and Kumar 2000) and the geographic distance matrix among the individuals analyzed. In this study, Mantel's statistic was normalized according to Smouse et al. (1986). This procedure transforms the statistic into a correlation coefficient. The geographic distances were measured with the Spuhler's (1972) procedure, where

$$D = \arcos \left(\cos X_{(i)} \cdot \cos X_{(j)} + \sin X_{(i)} \cdot \sin X_{(j)} \cos |Y_{(i)} - Y_{(j)}| \right),$$

where $X_{(n)}$ and $Y_{(n)}$ are the latitude and longitude of the nth individual sampled, respectively. The significance of the correlations obtained was tested using a Monte Carlo simulation with 5,000 permutations.

2. To determine possible isolation by distance among the haplotypes within the geographical area analyzed in *T. bairdii* and in *T. pinchaque*, the IBD version 1.2 software (Bohonak 2002) was employed. In this analysis, we used the quoted genetic distance against the geographical distance among the individuals sampled. The intercept and the slope of this relationship was calculated using Reduced Major Axis (RMA) regression (Sokal and Rohlf 1981; Hellberg 1994). Ten thousand randomizations (jackknife over population pairs and bootstrapping over independent population pairs) were executed to determine 95 and 99 % confidence intervals. The calculations were completed with non-transformed data and with log transformed data (genetic distance & geographical distance) jointly and separately.

3. A spatial autocorrelation analysis (Sokal and Oden 1978ab; Sokal and Wartenberg 1983; Sokal et al. 1986, 1987, 1989; Sokal and Jacquez 1991; Epperson 1990, 1993; Ruiz-García 1998, 1999 and Ruiz-García and Jordana 1997, 2000) was applied to the different haplotypes found in both species (separately, of course). The most frequent haplotype was weighted as 1, while the rest of the haplotypes were differentially weighted depending on the number of nucleotide substitutions differing from the most frequent haplotype. Autocorrelation coefficients and correlograms were estimated. For this, the Moran's I index and the Geary's c coefficient (Moran 1950; Durbin and Watson 1950) were employed

in the current study. In the case of *T. bairdii*, three distance classes (DC) were defined (3 DC: 0-161 km; 161-776 km; 776-1,893 km), while in the case of *T. pinchaque*, four DC were defined (4 DC: 0-71 km; 71-163 km; 163-494 km; 494-772 km). The criteria used, to choose these particular distance classes, was a relatively equal number of point pairs, among distance classes. To use these statistics, individuals must be connected using some type of network, which simulates as realistically as possible, the relationships existing between them. In this case, three network connections were used. The first method was binary, with all pairs of individuals connected at different specified distance classes. These connections were determined by using the possible gene flow routes between the individuals considered (Trexler 1988). Also, the Gabriel-Sokal network (Gabriel and Sokal 1969; Matula and Sokal, 1980) and the Delaunay's triangulation with elimination of the crossing edges (Ripley 1981; Upton and Fingleton 1985; Isaaks and Srivastava, 1989) were used. However, the results were very similar in each case. The Bonferroni (Oden 1984), Oden's Q and the Kooijman's tests were calculated with SAAP 4.3 software to determine the statistical significance of autocorrelation.

Another different spatial autocorrelation analysis was carried out. In this case, for *T. bairdii*, seven variables (the polymorphic nucleotide sites) were employed, while 47 variables were used for *T. pinchaque*. Distograms (Degen and Scholz 1998; Vendramin et al. 1999) and correlograms were estimated in both cases among individuals. For the distograms, two procedures were employed: the Gregorious's (1978) genetic distance and the number of common haplotypes (Hamrick et al. 1993). For the correlograms, the Moran´s I index and the Geary´s c coefficient were employed as before. Three and four DC were employed in both *Tapirus* species. In this case, the significance of distograms, correlograms and autocorrelation coefficients was calculated by means of 1,000 Monte-Carlo simulations (Manly 1997) and 95 % confidence intervals were estimated (Streiff et al. 1998). Also, the Bonferroni procedure was employed to determine the significance of these autocorrelation coefficients. For this analysis, the SGS version 1.0d software was applied (Degen et al. 2001).

3. Results

3.1 Gene diversity, genetic heterogeneity and demographic changes in the three Neotropical tapirs

The mutation model, which best fitted for all the *Tapirus* sequences, was the Jukes-Cantor model if we consider the AIC criteria (AIC = 8.3177) or the GTR model of nucleotide substitution with gamma distributed rate variation among sites if we consider the maximum likelihood criteria (LnL = -3.2603).

Of the three *Tapirus* species studied, *T. terrestris* clearly showed the highest levels of genetic diversity (Table 1): S = 107, H = 80, H_d = 0.984 ± 0.003, π = 0.0114 ± 0.0003, k = 10.335 ± 4.743 and $\theta_{per\ gene}$ = 19.376 ± 4.739. *T. pinchaque* showed the second level of gene diversity, while *T. bairdii* showed the lowest. Therefore, taking into account some relative gene diversity statistics, such as π, that are not affected by the sample size, *T. terrestris* presented 1.46 times more genetic diversity than *T. pinchaque* and 4.56 times more genetic diversity than *T. bairdii*. *T. pinchaque* showed 3.12 times more genetic diversity than *T. bairdii*.

	S	NH	Hd	π	K	θ per sequence
Tapirus terrestris	107	80	0.984 ±0.003	0.0114 ±0.0003	10.335 ±4.743	19.376 ±4.739
Tapirus pinchaque	47	10	0.895 ±0.070	0.0078 ±0.0046	7.029 ±4.498	14.455 ±5.565
Tapirus bairdii	7	6	0.800 ±0.114	0.0025 ±0.0005	2.182 ±1.306	2.390 ±1.251

Table 1. Genetic diversity statistics estimated for the three *Tapirus* species studied (*T. terrestris, T. pinchaque, T. bairdii*). S = number of polymorphic sites; NH = Number of Haplotypes determined; π = Nucleotide diversity; K = Average number of different nucleotides within each group analyzed; θ per sequence (= $2N_e\mu$), being N_e, the effective female population size, and μ, the mutation rate per generation.

The genetic divergence among the three neotropical *Tapirus* species by means of the mtCyt-b gene was highly significant (Table 2). For instance, the γ_{st}, N_{st} and F_{st} statistics yielded extreme genetic differentiation (0.789, 0.916 and 0.911, respectively) among the three species considered, with virtually no gene flow among them (Nm = 0.13, 0.05, 0.05, respectively). When the genetic heterogeneity was estimated by species pairs (Table 3), it was observable that the genetic differentiation of *T. bairdii* was noteworthy higher with regard to the two South-American species (F_{st} = 0.952 respect to *T. pinchaque* and F_{st} = 0.941 respect to *T. terrestris*), whereas between the two South-American species, the genetic differentiation was considerably lower (F_{st} = 0.491).

Genetic differentiation estimated considering the three *Tapirus* species analyzed	P	Gene flow	
χ^2 = 132.000 df = 88	0.0017*		
H_{ST} = 0.0485	0.0000*	γ_{ST} = 0.7898	Nm = 0.13
K_{ST} = 0.7851	0.0000*	N_{ST} = 0.9159	Nm = 0.05
K_{ST}^* = 0.3789	0.0000*	F_{ST} = 0.9108	Nm = 0.05
Z_S = 462.7719	0.0000*		
Z_S^* = 5.6642	0.0000*		
S_{nn} = 0.9849	0.0000*		

Table 2. Diverse genetic heterogeneity statistics (χ^2, H_{ST}, K_{ST}, K_{ST}^*, Z_S, Z_S^*, S_{nn}) and their associated probabilities taken together the three *Tapirus* species analyzed (*T. terrestris, T. pinchaque, T. bairdii*). All the probability values were significant (* P < 0.00001). Also some gene flow estimates (Nm) and the genetic heterogeneity statistics from which derived are shown. All the Nm estimates were lower than 1 indicating that genetic isolation exists between the three *Tapirus* species analyzed (*T. terrestris, T. pinchaque, T. bairdii*).

	F_{ST}	γ_{ST}	1	2	3
1 *Tapirus terrestris*				0.280	0.803
2 *Tapirus pinchaque*	0.491				0.906
3 *Tapirus bairdii*	0.941	0.952			

Table 3. Genetic heterogeneity statistics (F_{ST}, γ_{ST}) among all the pairs of the different *Tapirus* species considered.

The results of the historical demographic analyses are provided in Figure 3 and Table 4. *T. terrestris* was the species which undoubtedly crossed a population expansion during its natural history. The mismatch distribution, as well as all the statistics applied, showed significant evidence of population expansion for this species. In the case of *T. pinchaque*, the evidence of population expansion is relatively weaker. Only three statistics showed significant evidence of a possible population expansion (Fu and Li D*, Fu and Li F* and Tajima D statistics), but the mismatch distribution analysis (and the *rg* statistic) as well as the Fu's F_s and the R2 statistics (the two last being the most powerful tests to detect demographic changes) did not reveal any significant trend. In the case of *T. bairdii*, the situation is no clear. The mismatch distribution (and the *rg* statistic) revealed a significant population expansion, but neither of the other statistics revealed a significant population change trend. Therefore, only *T. terrestris* showed an undisputable pattern of population expansion, whereas the other two *Tapirus* species showed inconclusive results with regard to this topic.

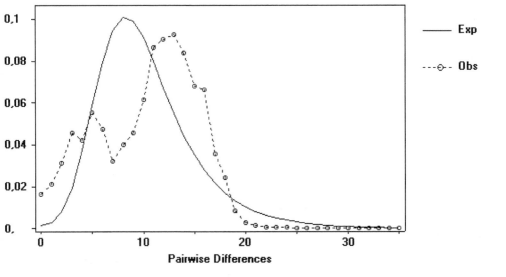

Tapirus terrestris

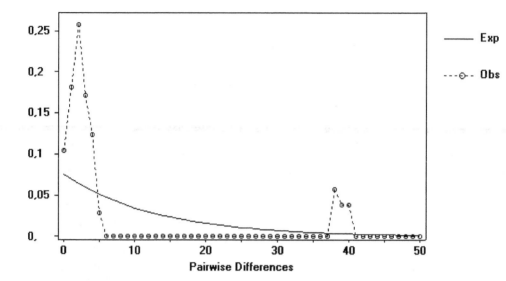

Tapirus pinchaque

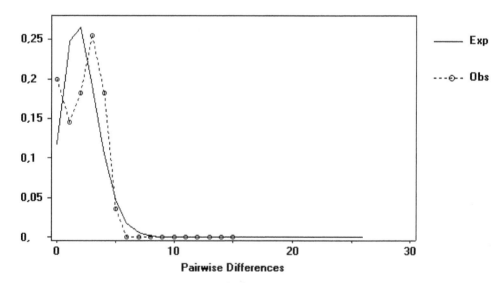

Tapirus bairdii

Fig. 3. Historical demographic analyses by means of the mismatch distribution procedure (pairwise sequence differences) with mtCyt-b gene sequences in the three *Tapirus* species studied.

	Tajima D	Fu & Li D*	Fu & Li F*	Fu's Fs	raggedness rg	R2
Tapirus terrestris	P[D ≤ -1.642] = 0.018*	P[D* ≤ -4.00] = 0.004**	P[F* ≤ -3.59] = 0.004**	P[Fs ≤ -34.02] = 0.000**	P[rg ≤ 0.0035] = 0.0009**	P[R2 ≤ 0.0453] = 0.031*
Tapirus pinchaque	P[D ≤ -2.207] = 0.002**	P[D* ≤ -2.843] = 0.004**	P[F* ≤ -3.071] = 0.003**	P[Fs ≤ -1.028] = 0.302	P[rg ≤ 0.0362] = 0.195	P[R2 ≤ 0.1955] = 0.932
Tapirus bairdii	P[D ≤ 0.353] = 0.384	P[D* ≤ -0.193] = 0.402	P[F* ≤ 0.263] = 0.389	P[Fs ≤ -1.331] = 0.146	P[rg ≤ 0.0374] = 0.024*	P[R2 ≤ 0.1331] = 0.111

Table 4. Demographic statistics applied to the three neotropical *Tapirus* species studied. * P < 0.05; ** P < 0.01, significant population expansions.

3.2 Phylogenetics, temporal splits and phylogeography in *T. pinchaque* and *T. bairdii*

The phylogenetic relationships and the possible temporal splits among the four *Tapirus* species were analyzed by means of the BEAST v. 1.4.8 software (Figure 4). All four *Tapirus* species were monophyletic, with the first split between *T. indicus* and the three neotropical *Tapirus* species, the second split between *T. bairdii* and the clade *T. terrestris-T. pinchaque* and the third split between *T. terrestris* and *T. pinchaque*. All the probabilities of the main clades of each one of the *Tapirus* species are equal or almost equal to 1.

This analysis offered the following temporal separation estimations: the ancestor of *T. indicus* diverged around 17 MYA (95 % HPD: 15.1-19 MYA) from the ancestor of the three neotropical tapirs, while the ancestor of *T. bairdii* diverged around 10.9 MYA (95 % HPD: 5.3-16.3 MYA) from the *T. terrestris-T. pinchaque* clade and the ancestors of *T. terrestris* and *T. pinchaque* diverged around 3.8 MYA (95 % HPD: 3.1-4.7 MYA). Additionally, the temporal diversification of the current *T. bairdii* began around 2.4 MYA (95 % HPD: 0.8-4.5 MYA), the diversification of the current *T. terrestris* around 3.5 MYA (95 % HPD: 2.3-4.4 MYA) and the diversification of *T. pinchaque* around 2.1 YA (95 % HPD: 1-3.3 MYA).

The mutation rate was around 3.9×10^{-3} during the separation of the ancestors of the Asian tapir and the neotropical ones and during the separation of the ancestor of the Central-American tapir against the ancestor of the South-American tapir. This mutation rate accelerated nearly three times during the separation of the ancestors of *T. terrestris* and *T. pinchaque* (1.03×10^{-2}) and in the initial lineage diversification within *T. bairdii* (1.01×10^{-2}). However, within the diversification of the *T. terrestris* lineages and within the diversification of *T. pinchaque* lineages (and also in recent *T. bairdii* lineages), the mutation rates were similar to those of the initial separation branches. Therefore, a period of mutation acceleration seems to have occurred during the ancestral divergence between the two South-America tapir species and during the initial lineage diversification within *T. bairdii*.

The time splits within *T. pinchaque* and *T. bairdii* are as follows: 1- As it was mentioned earlier, the ancestor of the *T. pinchaque* haplotypes began to diversify 2.11 MYA. Two

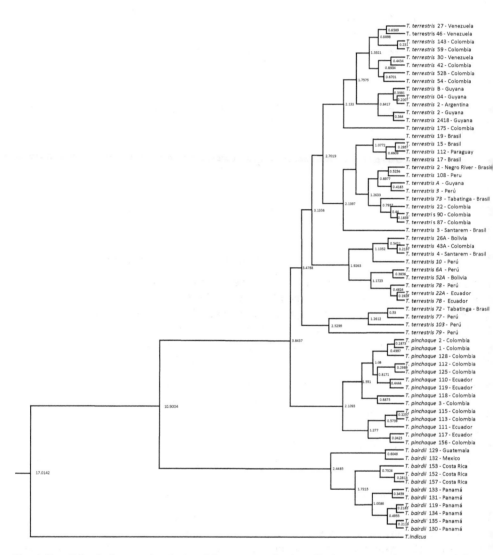

Fig. 4. Possible phylogenetic relationships among the four *Tapirus* species by means of the BEAST v. 1.4.8 software.

lineages were generated, one, with an ancestor around 1.51 MYA and other with an ancestor around 1.28 MYA. Both lineages were inter-dispersed within Colombia and Ecuador. The first lineage was subdivided in other two ensembles. One of them began its diversification around 1.08 MYA and the other around 0.89 MYA. This last one was composed of haplotypes only found in Colombia. The first one was comprised of haplotypes simultaneously found in Colombia and Ecuador and that diverged 0.87, 0.50, 0.44, 0.30 and 0.29 MYA. The other main *T. pinchaque* lineage had an ancestor which began to diversify

.28 MYA and internal ensembles that began to diversify around 0.57, 0.42 and 0.23 MYA. 2-
\s commented before, the ancestor of the *T. bairdii* haplotypes began its diversification 2.44
/IYA and determined one lineage, which is found in Mexico and Guatemala (and that
nternally diversified around 0.60 MYA in the northern area of the distribution), and another
ineage, in the southern range of the distribution, whose diversification began 1.72 MYA,
)riginating the animals from Costa Rica (with a common ancestor 0.75 MYA and with
ıdditional haplotype diversification 0.28 MYA) and the animals from Panama, with a
:ommon ancestor 1 MYA. Within the Panamanian animals, different haplotype split times
vere 0.50, 0.35 and 0.22-0.21 MYA.

3y employing another approximation (MJ with the ρ statistic), the following temporal splits
vere estimated (Figure 5): 1- The divergence between the haplotype of *T. indicus* and two of
he main *T. terrestris*'s haplotypes were 19.037 ± 0.282 MYA and 17.542 ± 0.153 MYA, or
)etween the first haplotype and the more frequent one in *T. pinchaque* was 19.830 ± 0.448
/IYA, very similar to the split time estimated for the initial divergence time between the
\sian tapir and the Neotropical tapir species with the best Bayesian hypothesis tree. The
wo most frequent haplotypes of *T. terrestris* diverged from the ancestor of the current *T.*
)airdii since 9.582 ± 0.157 MYA and 7.931 ± 0.076 MYA (and this divergence estimate was of
).559 ± 0.102 between the ancestors of *T. bairdii* and *T. pinchaque*), while these same two *T.*
errestris*'s haplotypes diverged from the main *T. pinchaque*'s haplotype 1.582 ± 0.299 MYA
ınd 1.525 ± 0.336 MYA, respectively. 2- Within *T. bairdii*, some interesting split divergence
imes were as follows: the ancestor of the main Panamanian's haplotype diverged from the
ıncestors of the main Costa Rican, Guatemalan and Mexican haplotypes 33,897 ± 33,897 YA,
116,219 ± 58,110 YA and 135,589 ± 33,897 YA while the two Costa Rican's haplotypes
diverged 67,795 ± 67,795 YA and the Guatemalan and the Mexican's haplotypes diverged
(01,692 ± 19,692 YA. 3- Within *T. pinchaque*, the main Tolima's haplotype separated from the
najor part of the other haplotypes found in Colombia and Ecuador around 65,627 ± 32,813

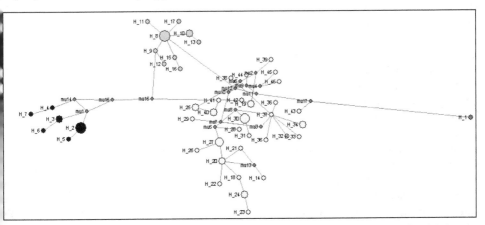

Fig. 5. Median Joining Network (MJ) applied by means of the software Network 4.2.0.1
Fluxus Technology Ltd) to the haplotypes found in four *Tapirus* species. In blue = *Tapirus*
ndicus; In yellow = *Tapirus terrestris*; In green = *Tapirus pinchaque*; In black = *Tapirus bairdii*.
In red, those hypothetical connecting haplotypes that were not detected in our study.

YA, with the lowest value around 32,813 ± 32,813 YA with another Ecuadorian's haplotype and the highest value around 98,440 ± 46,405 YA with Huila's haplotype. In whatever case, all the haplotype splits within *T. bairdii* and *T. pinchaque* were during the last phase of the Quaternary, mainly since the last glacial period to the beginning of the Holocene.

3.3 Spatial genetic patterns in *T. pinchaque* and in *T. bairdii*

The first spatial analysis revealed that *T. pinchaque* did not show any significant spatial trend in the distribution of its haplotypes (r = -0.119, approximate Mantel t-test = -0.660, p = 0.254; out of 5000 random permutations, 2112 were < Z, 0 = Z and 2888 > Z; one-tail probability p = 0.423). On the contrary, the *T. bairdii*'s haplotypes showed a very significant spatial genetic trend (r = 0.819, approximate Mantel t-test = 3.611, p = 0.0002; out of 5000 random permutations, 5000 were < Z, 0 = Z and 0 > Z; one-tail probability p = 0.0004).

The second analysis by means of the IBD software confirmed that *T. pinchaque*'s haplotype distribution did not follow an isolation by distance model [lineal model: intercept (± standard error) = 0.02587 ± 0.00227, slope (± standard error) = -0.007058 ± 0.000691, R^2 = 0.0142; 99% confidence intervals with 10,000 bootstraps over all the individuals: intercept = (-0.01207, 0.03452), slope = (-0.009161, -0.008307), R^2 = (0.0000, 0.0985)], while the individual spatial distribution of the *T. bairdii*'s haplotypes clearly fit with an isolation by distance pattern [lineal model: intercept = 0.000636 ± 0.000199, slope = 0.0004294 ± 0.0000338, R^2 = 0.670; 99% confidence intervals with 10,000 bootstraps over all the individuals: intercept = (0.0002354, 0.001088), slope = (0.0003631, 0.0005143), R^2 = (0.454, 0.820)]. These results were obtained with normal data, but the results with the log genetic or log geographical distances or both log data simultaneously were very similar.

The first spatial autocorrelation analysis (Table 5) did not show any spatial trend in *T. pinchaque*. Neither the overall correlogram (p = 0.455, for the Morans's I index; p = 0.521, for the Geary´s c coefficient) nor any individual autocorrelation coefficient were significant. Even, the most negative value for the Moran´s I index of the four defined distance classes (DC) was the first one (-0.20) and the most positive was the fourth DC (0.02). Therefore, this means that the most differentiated haplotypes were those geographically closest, whereas the more similar haplotypes were at greater geographical distances. Contrarily, this first spatial autocorrelation analysis showed a striking significant result for the overall correlogram in *T. bairdii* related to the monotonic clinal pattern for both the Moran's I index and the Geary's c coefficient (Table 5), with the first DC highly positive (Moran's I index 1 DC = 0.500, p = 0.000; Geary's c coefficient 1 DC = 0.05, p = 0.000) and the third DC highly negative (Moran's I index 3 DC = -0.68, p = 0.000; Geary's c coefficient 3 DC = 2.29, p = 0.000). Similarly, the same was found for *T. pinchaque* with no spatial autocorrelation (Table 5) with the distogram with the Gregorious (1978)'s distance or the distogram with the number of haplotypes shared (SGS software). Contrarily, again, in *T. bairdii* the spatial patterns were highly significant for the distogram with the Gregorious (1978)'s distance or the distogram with the number of haplotypes shared.

Moran´s I index	1CD	2CD	3CD	4CD	Overall Probability Correlogram
	0-71 km	71-163 km	163-494 km	494-772 km	
Tapirus pinchaque	-0.20	-0.03	-0.08	0.02	0.455
	0-161 km	161-776 km	776-1893 km		
Tapirus bairdii	0.50**	-0.09	-0.68**		0.000**

A

Geary's c coefficient	1CD	2CD	3CD	4CD	Overall Probability Correlogram
	0-71 km	71-163 km	163-494 km	494-772 km	
Tapirus pinchaque	1.74	0.89	0.85	0.53	0.521
	0-161 km	161-776 km	776-1893 km		
Tapirus bairdii	0.05**	0.59	2.29**		0.001**

A

Distogram with Gregorious (1978)'s distance	1CD	2CD	3CD	4CD	Overall Probability Correlogram
	0-1.9175 units	1.9175-3.835 units	3.835-5.7525 units	5.7525-7.67 units	
Tapirus pinchaque	0.1713 (95% interval: 0.0691-0.1878)	0.0745 (95% interval: 0.0213-0.7021)	0.1654 (95% interval: 0.0696-0.4004)	0.0940 (95% interval: 0.0359-0.4199)	1 CD: (95% interval: 0.494-0.506) 2 CD: (95% interval: 0.497-0.503) 3 CD: (95% interval: 0.221-0.779) 4 CD: (95% interval: 0.700-0.300)
	0-4.37 units	4.37-8.74 units	8.74-13.110 units	13.11-17.48 units	
Tapirus bairdii	0.0601 (95% interval: 0.1955-0.3759)	0.3571 (95% interval: 0.2142-0.4365)	0.5476 (95% interval: 0.1429-0.5000)	0.5195 (95% interval: 0.1688-0.5195)	1 CD: (95% interval: 1.000-0.000)** 2 CD: (95% interval: 0.180-0.820) 3 CD: (95% interval: 0.020-0.980)* 4 CD: (95% interval: 0.017-0.983)*

B

Table 5. A- Spatial autocorrelation analysis with 4 Distance Classes (DC) for the individuals analyzed of *Tapirus pinchaque* and with 3 DC for the individuals analyzed of *Tapirus bairdii*. * $P < 0.05$, ** $P < 0.01$, significant probability. This analysis was carried out with the SAAP 4.3 program. B- Distograms with the Gregorious's (1978) distance carried out with the SGS software for the individuals analyzed of *Tapirus pinchaque* and *Tapirus bairdii*. * $P < 0.05$, ** $P < 0.01$, significant probability.

4. Discussion

4.1 Genetic diversity, genetic heterogeneity, possible demographic changes and spatial structure

It was clear that *T. terrestris* was the species with the highest gene diversity level, which agrees quite well with the fact that it is the Latin American *Tapirus* species with the widest geographical distribution and thus with the highest potentially effective numbers. This coincides with the elevated polymorphism found for this species with other markers (Tapia et al. 2005). Curiously, *T. pinchaque*, a species with a very restrictive geographic distribution and a small population size of no more than a thousand individuals, which could be on the brink of extinction (Ashley et al. 1996), presented a gene diversity that was more than three times higher than that of *T. bairdii* which historically occupied a distribution from southern Mexico to the Pacific area of Colombia and Ecuador (historically in Ecuador until the Guayaquil Gulf and currently in Guayas-Bucay, Cotacachi-Cayapas Reserve and Awá Reserve; Tirira 2008). Thus, although *T. pinchaque* has a very small census population size and a very restrictive geographical distribution, within disturbed and fragmented ecosystems, the species is not depauperated from a genetics point of view. This is a good new for conservation purposes. However, this is the first genetics analysis reported for this species and other genetic markers, such as DNA microsatellites, must be analyzed for this elusive species to corroborate that its genetic diversity levels are high although it has an extremely small population size. If this affirmation is correct, this means that the reproductive system of this species, and its gene flow capacity dispersion, is enough to maintain these elevated gene diversity levels. The case of the *T. bairdii* seems to be more dramatic. Although its geographic distribution is wide, it has been dramatically reduced and fragmented in the last two centuries and today no more than 6,000 individuals are left in the wild (Ashley et al. 1996). Its mitochondrial genetic diversity levels were extremely low compared with the other two neotropical tapirs and other neotropical mammals studied for the same or similar mitochondrial genes (Primates, Lavergne et al. 2010; Ruiz-García and Pinedo-Castro 2010; Ruiz-García et al. 2010, 2011a,b, 2012a; jaguars, pumas and other felids, Ruiz-García et al. 2006, 2009a, Cossios et al. 2009; artiodactyls, Ruiz-García et al. 2007, 2009b; river dolphins, Banguera-Hinestroza et al. 2002; Ruiz-García 2010a,b; Ruiz-García et al. 2008; Caballero et al. 2010). This could mean that this species suffered from a bottleneck and/or the gene drift has been more intense on this species by natural or human constrictions. In fact, this species has intensely declined in the last century by habitat destruction and hunting and has been extinct in El Salvador and in a major fraction of its original distribution range in Colombian and Ecuador (Brooks et al. 1997). Additionally, this species also showed a low gene diversity level for DNA microsatellite markers, with an expected heterozygosity ranging from 0.37 (Costa Rica sample) to 0.43 (Panama sample) and the average number of alleles oscillating from 2.5 (Costa Rica) to 3.33 (Panama) (Norton and Ashley 2004a). These microsatellite gene diversity levels were among the lowest found for mammals and similar to other genetically depauperated neotropical mammals, such as the Andean bear (Ruiz-García 2003, 2007, 2012; Ruiz-García et al. 2005) or the Andean cat (Cossios et al. 2012). These levels, together with the low mitochondrial gene diversity herein reported, agree quite well with a history of isolation, gene drift or bottlenecks for this population. Nei et al. (1975) showed that the populations which did not quickly recover their population sizes following a bottleneck will experience a greater loss of the gene diversity levels and will take longer to recover heterozygosity (or in this case, nucleotide diversity). Moreover, the mutation process to add new genetic variants could take thousands of generations resulting in the maintenance of low allele or haplotypic diversity for thousands of years following the original bottleneck. Norton and Ashley (2004a) did not find clear evidence of bottlenecks in the Costa Rican Corcovado National Park *T. bairdii* population nor in the Panamanian population because they did not observe a heterozygosity excess expected in populations experiencing a recent bottleneck. Nevertheless, in the Costa Rican population, they determined an allele frequency mode shift which is consistent with a bottleneck. In our case, the mtCyt- gene sequences yielded, for *T. bairdii*, the weakest evidence of a possible population expansion relative to the cases of the other two neotropical tapir species, which is concordant with the fact that this is probably the most bottlenecked species. Contrarily, *T. terrestris* experienced a clear historical population expansion during the Pleistocene, becoming probably the most successful large herbivore in South America, which survived the last glacial extinction. In the case of *T. pinchaque*, there is less conclusive evidence of population expansion and it seems to have a constant population size throughout its history.

The genetic heterogeneity among the three *Tapirus* species was highly significant, showing that, effectively, they were three separated species. However, the two South-American species were highly related, while the Central American species was considerably divergent from the two South-American ones. This agree quite well with the morphological classification of Hershkovitz (1954), who had recognized the two South-American tapir species as belonging to the sub-genus *Tapirus*, while the *T. bairdii* was located in the sub-genus *Tapirella*. This is clear evidence that in the Neotropics, there are currently two different molecular *Tapirus* lineages.

Other results which should be commented upon are the absence of any spatial pattern in the case of *T. pinchaque* and the striking spatial pattern discovered in *T. bairdii*. Although the current geographical distribution of *T. pinchaque* is restrictive and the populations could be severely fragmented, the intense climatic changes in the Andes during the Pleistocene probably provoked the stochastic mixing of different haplotypes that were generated in diverse areas when the conditions were sufficiently adequate for population expansion. Additionally, the population size of this species could be higher in the past than it is currently, although several Andean mammal species have shown to have small effective numbers throughout their histories (Andean bear, Ruiz-García 2003, 2007, 2012; Ruiz-García et al. 2003, 2005; Andean cat, Cossios et al. 2012). Also, the females of *T. pinchaque* could have a great capacity for dispersion, which did not provide for the generation of an appreciable spatial pattern during the history of this species. In contrast, in the case of *T. bairdii*, the geographical distribution, and the minor extension, of the Central American forests could restrict the dispersion of this large herbivorous mammal compared to the South American forests. Thus, an isolation-by-distance pattern has been generated for *T. bairdii*. Norman and Ashley (2004a), by using microsatellites, did find an elevated F_{ST} value between Panamanian and Costa Rican tapirs, although they were not differentiated in two different populations by the Structure software. The results obtained seem to put forward the possible existence of Management Units (MU) in the case of *T. bairdii* but not for *T.*

pinchaque. One aspect which should be analyzed is if this significant spatial pattern and the limited gene diversity for mt sequences (this study) and for DNA microsatellites (Norton and Ashley, 2004a) is the result of the Pleistocene history or of the more recent history of this species.

4.2 Temporal splits within *T. pinchaque* and *T. bairdii*, climatic and geological changes and phylogenetics relationships among the Neotropical tapirs

Tapirus has been in South America since the lower Pleistocene, or Plio-Pleistocene, (around 3 MYA; Cione and Tonni 1996; Nabel et al. 2000) in the Ensenadan South American Land Mammal Age (SALMA), during the Great American Biotic Interchange (GABI). Our results showed that the ancestor of *T. terrestris* and *T. pinchaque* lived 3.8 MYA and the haplotypic diversification within *T. terrestris* and within *T. pinchaque* occurred 3.5 MYA and 2.1 MYA respectively. Thus, the diversification of the ancestors of the current South-America *Tapirus* coincides with the climatic changes that originated the completion of the Panamanian land bridge (2.8-3.5 MYA; Coates and Obando, 1996) or slightly earlier coinciding with the Chocó-Panamá island bridge (Galvis 1980), which could have been used by the ancestors of the current *Tapirus* to colonize northern South America from Central America. During the upper Pliocene orogeny, the present Tuira, Atrato and Sinú river basins as well as near lowlands were raised above sea level. Thus, the mountains of southern Central America and of the northern Andes were uplifted to about their present elevation (Van der Hammen 1961). Even if the divergence split was slightly prior to the completion of the Panamanian land bridge, and although the Nicaraguan, Panamanian and Colombian portals remained open (upper Miocene-Middle Pliocene), numerous volcanic island existed from the lower Atrato Valley and the Tuira river basin of eastern Panama to the Nicaraguan portal, which could have been used by the South-American *Tapirus*'s ancestor to migrate southward. The Cuchillo bridge of the Urabá region, connecting the Tertiary Western Colombian Andes with the Panamanian islands was probably above sea level during this period. Simpson (1950, 1965) claimed that many mammals were "island hoppers". Tapirs have a high capacity to swim wide zones of the Amazon River (Brooks et al. 1997).

The haplotypic time divergence within *T. terrestris* began around 3.5 MYA, when the *Tapirus* genus penetrated in South-America. It's likely that the *T. terrestris*' ancestor generated the ancestor of *T. pinchaque*, whose haplotypic diversification began more recently (2.1 MYA) than that of *T. terrestris*. Therefore, *T. pinchaque* could be a mountain specialized descendent of some *T. terrestris* lineage in the transition of the western Amazon and the Eastern Andes Cordillera. In fact, the northern Andes, where the current *T. pinchaque* lives, are a hot point of speciation for taxa (Sedano and Burns 2010) and a top biodiversity hotspot (Orme et al. 2005). Furthermore, the haplotype diversification of *T. pinchaque* began after the completion of the northern Andean uplift. Effectively, after rising slowly for millions of years, the Central Andes had a rapid and final uplift in the last 6-10 MYA (Garzione et al. 2008). The final uplift of the northern Andes in the Eastern Cordillera in Colombia was in the last 3-6 MYA (Hooghiemstra and Van der Hammen 2004), while the beginning of the haplotype diversification in *T. pinchaque* was around 2.1 MYA. This molecular result did not agree with the fact that *T. pinchaque* could be more primitive and to be in origin of *T. terrestris* as certain authors claimed (Hershkovitz 1954).

n the evolution of *T. pinchaque* some ages seems to be especially important for haplotypic liversification. These temporal haplotypic splits were around 2.1-1.5, 0.85-1.3, 0.4-0.6 and .2-0.3 MYA. The ages between 2.1 and 1.5 MYA coincide with the end of the Pliocene and he beginning of the Pleistocene (1.6-2.5 MYA). Therefore, the initial Pleistocene changes ould have generated the fragmentation of the *T. pinchaque* population. During that epoch, he Andean forests (where *T. pinchaque* lives) were transformed into open cold dry savannah 'paramo"), which could have potentially isolated ancestor populations of *T. pinchaque* in :olombia. This was also the epoch of the last upheaval of the central and northern Andes. 'or instance, during this time in Huancho (north of Lima-Peru) little lakes and the 'entanilla bay were formed. Also the Andean cordillera between Cajamarca and luancavelina (Peru) was created by volcanism in this period of intense climatic and ,eographic changes. Van der Hammen (1992) demonstrated that the mean temperature in he Colombian Andes was 4 °C lower than today. He also stated that the rain level was ower than the 500-1,000 mm reported for today. This epoch was characterized by a great auna change with the beginning of the Marplatense epoch in Argentina or the 'illafranquense epoch in Europe.

'he Milankovitch's cycles (each 19,000-24,000, 43,000 or 90,000-100,000 years) occurred across he Quaternary with its cold and dry phases and generated forest refuges in South America Haffer 1997; Whitmore and Prance 1987). Therefore, the Pleistocene forest refugia invoked by laffer (1969, 1982) could be very important for understanding the evolution of the three urrent Latin American *Tapirus* species. Environmental conditions in Central and South ∖merica were influenced by these alternating dry and wet climatic periods and by sea level luctuations. During the glacial periods, sea levels were lower by about 100 m, while sea level ose by about 30 to 50 m above the present level during the interglacials (Haffer 1967).

'he next haplotype diversification peak occurred around 0.8-1.1 MYA. The Pre-Pastonian ,lacial period (0.80-1.30 MYA), which was the highest glacial peak of the first Quaternary ,laciation (Nebraska-Günz), could have separated to a certain degree, some of these *T. inchaque*'s haplotypes. Another mammal, the Pampas cat (*Leopardus colocolo*) (Cossios et al. .009), suffered an intense genetic population fragmentation in this same epoch.

'he *T. pinchaque* haplotype differentiation around 0.4-0.6 MYA agrees quite well with the oldest epoch of the Mindel-Kansas glacial period (Elster glacial period for Scandinavia, ∔onaerense period for Argentina and Kamasiense I for Eastern Africa; 0.3-0.6 MYA). Also, he *T. pinchaque* haplotype differentiation that occurred around 0.2-0.3 MYA coincides with he Riss I glacial period for central Europe, the Illinois glacial period for North America, the ∍aale for the Nordic glaciations and the Kanjeriense for Eastern Africa. Finally, other naplotype diversification periods were 98,000, 66,000 and 33,000 YA. These different ages orresponded to the last glacial period which originated 130,000 to until 10,000 YA. The :emiense interglacial period (130,000 to 80,000 YA), was characterized by high emperatures, high rain precipitations and extensive forests of *Aliso, Vallea* and *Weinmannia* Van der Hammen 1992). However, this epoch had some short but very intense cold periods one event 95,000-100,000 YA), where the Andean forests disappeared around the :undiboyacense highlands in Colombia. These brief, intensely cold periods were recorded ∍y oxygen isotopes in Greenland's ice. Agreeing quite well with the haplotype liversification that occurred 66,000 YA, the first big cold period began during the last glacial epoch (Earlier Pleni-glacial period) around 60,000-70,000 YA. Around 26,000-35,000 YA during the middle-upper Pleni-glacial period, the climate was extremely dry. For instance the Bogotá and the Fúquene lakes, in the Colombian highlands, disappeared (Van Geel and Van der Hammen 1973) and typical dry vegetation such as *Symplocos, Myrica, Myrsine* and *Alnus* appeared in the northern Andes (Van der Hammen 1980). During this epoch, the Amazon was colonized by dry vegetation such as *Ilex*. Van der Hammen (1992) demonstrated that the upper Pleni-glacial (30,000-16,500 YA), when the Amazon was characterized as having a dry- ambient climate, coincided with the most cold and dry period in the Andes (26,000-14,000 YA) as well as with the most extensive ice period in the Northern Hemisphere. Thus, another *T. pinchaque* haplotype period of diversification could exist at the beginning of this period.

For *T. bairdii*, the situation is similar to *T. pinchaque*. The first haplotypic divergence processes in *T. bairdii* (1.7-2.4 MYA) coincided with the beginning of the Pleistocene. A second period of haplotype diversification in this species was around 0.6-0.75 MYA. This period agrees quite well with a high glacial peak of the second glacial period (Kansas-Mindel), which could promote population differentiation within *T. bairdii*. A third moment of haplotype diversification was around 0.28 MYA. As in the case of *T. pinchaque*, this coincides with the Riss I glacial period for central Europe and with the Illinois glacial period for North-America. Also, in the last glacial period (130,000-10,000 YA), there was an intense haplotype diversification in *T. bairdii* (136,000, 116,000, 102,000, 68,000, 34,000 YA), coinciding with some of the cold peaks in the Eemiense interglacial period, the beginning of the Earlier Pleni-glacial period and with the beginning of the Upper Pleni-glacial period.

Thus, the climatic changes during the Pleistocene could be decisive in understanding the haplotype differentiation within *T. pinchaque* and *T. bairdii*.

All the phylogenetic analyses carried out with the mtCyt-b gene showed the close genetic relationship between both South-America species, *T. terrestris* and *T. pinchaque*, while the Central American species, *T. bairdii*, seems to belong to another *Tapirus* lineage, but more related with the other neotropical tapirs than to the Asian tapir. This result disagrees with that obtained by Ashley et al. (1996) and Norton and Ashley (2000) with the mtCOII gene, which showed a clade conformed by *T. terrestris* and *T. pinchaque* and another clade integrated by *T. bairdii* and *T. indicus* (with 55 % bootstrap support). Nevertheless, the results obtained by Norton and Ashley (2000) with the 12S rRNA gene supported monophyly of neotropical tapirs (83 % bootstrap), with the South-American tapirs and the Central American ones as the current forms of two different tapir lineages. When both genes were employed by Norton and Ashley (2000), the monophyly of the neotropical tapirs was also supported (62 % bootstrap). Thus, similar to the two sequence sets provided by Norton and Ashley (2000), the molecular results we show herein are closely related to the paleontological data of South-American tapirs.

The fossil remains of *Tapirus* in South America are scarce and fragmentary. This fragmentation is important because in many cases it makes it difficult to obtain taxonomic conclusions. But, all the taxa seem to be highly related and they could belong to a unique tapir colonization of South America, just as the genetics results seem to indicate. The most representative *Tapirus* fossil records in South America are as follows: Ameghino (1902) found the left mandible with three pre-molars of a *Tapirus* taxon larger than the current *T.*

terrestris from the lower Pleistocene of Tarija (Bolivia). He named this taxa as *T. tarijensis*. Rusconi (1928) and Cattoi (1951) described, from some teeth material, two supposed *Tapirus* species of the lower Pleistocene (Ensenadense age: 2-0.5 MYA) located in the south eastern section of the Buenos Aires Province. They named these taxa as *T. australis* and *T. dupuyi*. Nevertheless, Ubilla (1983) and Tonni (1992) lately considered that these materials really did not constitute new species (*Tapirus* sp.). Cattoi (1957) described another *Tapirus* form (*T. rioplatensis*), larger than *T. terrestris*, also from Ensenadense in the north-western area of the Buenos Aires Province. Tonni (1992) described a *T. terrestris's* mandible from the Lujanense age (upper Pleistocene Lujanense period to present, the last 130,000 YA) collected in the Colon Department of the Entre Rios Province, as being the last and most southern record for this species. Also, Noriega et al. (2004) and Ferrero et al. (2007) found fossil fragments of *T.* cf. *terrestris* (a right hemi-maxille) for the El Palmar Formation at the El Boyero locality (upper Pleistocene) from the Entre Rio Province. Ferrero and Noriega (2003, 2005, 2007) registered another *Tapirus* species from the analysis of a complete skull collected from outcrops of the Arroyo Feliciano Formation in the Diamante Department (Upper Pleistocene; Lujanense period) in the Argentine area of Mesopotamia. This skull was morphometrically compared, with a cladistic analysis, to several Tapiridae such as *Miotapirus, Paratapirus, Plesiotapirus*, five North American species (*T. veroensis, T. haysii, T. johnsoni, T. webbi* and *T. polkensis*) as well as the fourth current living *Tapirus* species. This taxon has been named *T. mesopotamicus*. This species seems to be closely related with *T. pinchaque*, and both species are related to *T. terrestris*. Therefore, four taxa of *Tapirus* have been determined for Argentina (Forasiepi et al. 2007): *T. terrestris, T. sp, T. mesopotamicus*, and *T. rioplatensis*. In whatever case, the existence of fossil records of tapirs in these Argentinean areas is associated with climatic conditions hotter and more humid than climatic conditions today. In Uruguay, Ubilla (1983) determined another *Tapirus* species for the Libertad Formation (Lower Pleistocene) in the Montevideo Department. This species was named *T. oliverasi* and had a larger size than *T. terrestris* but smaller than *T. rioplatensis*. Ubilla (1996) described *T. terrestris* fossils from the Tacuarembó and Salto Departments and other materials classified as *Tapirus* sp. Additionally, Ubilla and Rinderknecht (2006) found a complete skull from the Tacuarembó Department, which could be a new species. Therefore, the presence of *Tapirus* is associated with tropical forests and the extension and diversification of the genus during a great part of the Pleistocene was higher than it is today (especially during the Ensenadense period). It is in Brazil where more fossil *Tapirus* records have been found. Winge (1906) determined another species, *T. cristatellus*, from the fossils of Lagoa Santa, Minas Gerais state. This species has a skull and teeth greater than the current *T. terrestris*. Many fossil remains of *T. terrestris* or *T. sp.* have been described in Iraí (Rio Grande do Sul; Souza-Cunha 1959), in Arroio Touro Passo (Bombin 1976), Quaraí River (Oliveira 1992), in Bom Jardim (Pernambuco state; Rolim 1974), in Arroio Chuí (Soliani 1973) and in the Upper Jurua River (Acre state; Rancy 1981). An abundance of *T. terrestris* fossils have also been found in Areia Preta, Jacupiranga (Sao Paulo state; Paula-Couto 1980), Barauna (Rio Grande do Norte state; Porpino and Santos 2003), Cavernas do Bauxi (Mato Grosso; Hirooka 2003), Cavernas do Japones e Nascente do Formoso (Serra da Bodoquena, Mato Grosso do Sul; Salles et al. 2006) and the Cavernas do Vale do Rio Rocha (Gramados, Parana state; Sedor et al. 2005). Holanda et al. (2005) described some new findings in three Brazilian localities. The molar length and the width indices showed dimensions within the proportions of *T. terrestris*, with the exception for the PM1 (a specimen of the Rondonia State) and for the M2 and M3 (the second specimen from the Rio Grande do Sul State) which are more quadrangular like in *T. veroensis* and *T. haysii* (Plio-Pleistocene of North America). All these paleontological records were dated in less of 3.0 MYA. Recall that our genetics results pointed out that the *T. terrestris* diversification began around 3.5 MYA.

Very recently, Holanda and Couzzol (2006) described three specimens from the Pleistocene at Acre and Rondonia. Their morphometric analyses indicated two new species: a more robust form than the current *T. terrestris* (the Acre specimens) as well as a more gracile form (the Rondonia specimen). These fossils were dated to the middle Pleniglacial period, around 30,000-45,000 years ago. Thus, many different *Tapirus* species have existed in South America during the Pleistocene.

However, the discovery of 75 individuals of *T. polkensis* in the Gray Fossil Site in eastern Tennessee showed that a unique species was present but it had considerable intraspecific variation including development of the sagittal crest, outline shape of the nasals and the number and relative strength of lingual cusps on the P1 (Hulbert et al. 2009). This means that some of the quoted fossils found in South-America should belong to *T. terrestris*. If ancient DNA could be extracted from some of these more recent tapir fossil remains, it could be conclusive to demonstrate if only one *Tapirus* migration wave arrived to South-America and if they were really different species from *T. terrestris*.

In contrast, the morphometrics of the current *T. bairdii* seems to be more related with the North-American fossil tapirs (Ferrero and Noriega 2007) [*T. johnsoni, T. simpsoni, T. polkenis* and *T. webbi* for the Miocene; *T. merriami, T. haysii* from the Pliocene and these last two species plus *T. veroensis* for the Pleistocene; Hulbert 1995, 1999; Hulbert and Wallace 2005; Spassov and Ginsburg 1999; Tong 2005; Hulbert et al. 2009]. This agrees quite well with the molecular data showed here, which showed *T. bairdii* as belonging to a different lineage. However, our molecular study contradicts the morphometrics study of Hulbert (1995), which considered *T. terrestris* and *T. bairdii* to be the closest sister taxa. However, Ferrero and Noriega (2007) showed a close relationship among *T. bairdii* and *T. haysii* and *T. veroensis*. Thus, the molecular data agree quite well with the fossil knowledge that we have for the American tapirs. The first North American tapir was *T. johnsoni* from the Miocene of Nebraska with a divergence about 9-11 MYA (Colbert 2005). This temporal split agrees with the estimates obtained with the molecular data for the divergence between the lineage of *T. bairdii* and the lineage of *T. terrestris-T. pinchaque* (10.9 MYA for the Bayesian hypothesis tree and 9.6 MYA for the haplotype MJ network).

Future works should utilize nuclear sequences (introns, HLA loci, Chromosome Y) and microsatellites to study the phylogenetic relationships among the four living *Tapirus* species as well as the genetic structure for each one of the *Tapirus* species.

5. Acknowledgment

Thanks to Dr. Diana Alvarez, Pablo Escobar-Armel and Luisa Fernanda Castellanos-Mora for their respective help in obtaining *Tapirus terrestris* samples during the last 14 years. Many thanks go to the Peruvian Ministry of Environment, to the PRODUCE (Dirección Nacional de Extracción y Procesamiento Pesquero from Peru), Consejo Nacional del

Ambiente and the Instituto Nacional de Recursos Naturales (INRENA) and to the Colección Boliviana de Fauna (Dr. Julieta Vargas) for their role in facilitating the obtainment of the collection permits in Peru and Bolivia. The first author also acknowledges and thanks the Ticuna, Yucuna, Yaguas, Witoto and Cocama Indian communities in the Colombian Amazon, Bora, Ocaina, Shipibo-Comibo, Capanahua, Angoteros, Orejón, Yaguas, Cocama, Kishuarana and Alama in the Peruvian Amazon, to the Sirionó, Canichana, Cayubaba and Chacobo in the Bolivian Amazon and Marubos, Matis, Mayoruna, Kanaimari, Kulina, Maku and Waimiri-Atroari communities in the Brazilian Amazon for helping to obtain *Tapirus terrestris* samples. Thanks go to Dr. Luís Albuja for their help in obtaining *Tapirus pinchaque* samples from Ecuador.

5. References

Ameghino F, 1902. Notas sobre algunos mamíferos fósiles nuevos o poco conocidos del Valle de Tarija. Anales del Museo de Historia Natural de Buenos Aires 3: 225-261.

Ashley MV, Norman JE and Stross L, 1996. Phylogenetic analysis of the periossodactylan family Tapiridae using mitochondrial cytochrome c oxidase (COII) sequences. Journal of Mammalian Evolution 3: 315-326.

Bandelt H-J, Forster P and Rohl A, 1999. Median-joining networks for inferring intraspecific phylogenies. Mol Biol Evol 16: 37-48.

Banguera, E., Cardenas, H., Ruiz-García, M., Marmontel, M., Gaitán, E., Vásquez, R., García-Vallejo, F. 2002. Molecular Identification of Evolutionarily units in the Amazon River dolphin, *Inia sp* (Cetacea: Iniidae). Journal of Heredity 93: 312-322.

Bohonak AJ, 2002. IBD (Isolation by Distance): A program for Analyses of isolation by distance. Journal of Heredity 93: 153-154.

Bombin M, 1976. Modelo paleoecológico evolutivo para o Neoquaternário da regiao da Campanha-Oeste do Rio Grande do Sul (Brasil). A formacao Touro Passo, seu contéudo fossilífero e a pedogenese pós-deposicional. Comunicacoes do Museu de Ciencias da PUCRGS 15: 1-70

Brooks DM, Bodmer RE, Matola S, 1997. *Tapirs - Status Survey and Conservation Action Plan.* IUCN/SSC Tapir Specialist Group. IUCN, Gland, Switzerland and Cambridge, UK. viii + 164 pp.

Caballero S, Trujillo F, Ruiz-García M, Vianna J, Marmontel M, Santos FR, Baker CS, 2010. Population structure and phylogeography of tucuxi dolphins (*Sotalia fluviatilis*). In *Biology, Evolution, and Conservation of River Dolphins Within South America and Asia.* Ruiz-García, M., Shostell, J. (Eds) Nova Science Publishers., Inc. New York. Pp. 285-299.

Cattoi N, 1951. El status de *Tapirus dupuyi* (C. Amegh). Comunicaciones del Museo Argentino de Ciencias Naturales "Bernardino Rivadavia" e Instituto Nacional de Investigación de las Ciencias Naturales 2: 103-112.

Cattoi N, 1957. Una especie extinguida de *Tapirus* Brisson (*T. rioplatenses* nov. sp). Ameghiniana 1: 15-21.

Cione AL, Ton EP, 1996. Reassessment of the Pliocene-Pleistocene continental time scales of Southern South America. Correlations of the type Chapadmalalan with Bolivian sections. Journal of South American Earth Sciences 9: 221-236.

Coates AG, Obando JA, 1996. The geologic evolution of the Central American Isthmus. In Jackson, J. B. C., Budd, A. F., Coates, A. G. (Eds.). *Evolution and Environment in Tropical America*. The University of Chicago Press, Chicago. Pp. 21-56.

Colbert M, 2005. The facial skeleton of the Early Oligocene *Colodon* (Perissodactyla, Tapiroidea). Palaeontologia Electronica 8: 1-27.

Colbert M, 2007. New fossil discoveries and the history of *Tapirus*. Tapir Conservation-The Newsletter of the IUCN/SSC Tapir Specialist group- 16: 12-14.

Cortes-Ortiz L, Birmingham E, Rico C, Rodriguez-Luna E, Sampaio I, Ruiz-García M, 2003. Molecular systematics and biogeography of the Neotropical monkey genus *Alouatta*. Mol Phylogenet Evol 26: 64-81.

Cossios ED, Lucherini M, Ruiz-García M, Angers B, 2009. Influence of ancient glacial periods on the Andean fauna: the case of the Pampas cat (*Leopardus colocolo*). BMC Evolutionary Biology 9: 68-79.

Cossios ED, Walker S, Lucherini M, Ruiz-García M, Angers B, 2012. Between high-altitude islands and high-altitude corridors. The population structure of the Andean cat (*Leopardus jacobita*). Endangered Research Species (in press).

Degen B, Scholz F, 1998. Spatial genetic differentiation among populations of European beech (*Fagus sylvatica* L) in Western Germany as identified by geostatistical analysis. Forest Genetics 5: 191-199.

Degen B, Petit R, Kremer A. 2001. SGS -Spatial Genetic Software: A computer program for analysis of spatial genetic and phenotypic structures of individuals and populations. Journal of Heredity 92: 447-448.

Deng T. 2006. Paleoecological comparison between late Miocene localities of China and Greece based on *Hipparion* faunas. Geodiversitas 28: 499-516.

Drummond AJ and Rambaut A, 2007. BEAST: Bayesian evolutionary analysis by sampling trees. BMC Evol. Biol. 7: 214.

Drummond AJ, Ho SYW, Phillips MJ and Rambaut A, 2006. Relaxed phylogenetics and dating with confidence. PLOS Biol. 4(5): e88.

Durbin J and Watson GS, 1950. Testing for serial correlation in least squares regression. Biometrika 37: 409-428.

Epperson BK, 1990. Spatial autocorrelation of genotypes under directional selection. Genetics 124: 757-771.

Epperson BK, 1993. Recent advances in correlation studies of spatial patterns of genetic variation. Evolutionary Biology 27: 95-155.

Ferrero BS and Noriega JI, 2003. El registro fósil de los tapires (Perisodactyla: Tapiridae) en el Pleistoceno de Entre Ríos. Ameghiniana 40: 84R.

Ferrero BS and Noriega JI, 2005. Tapires del Pleistoceno de Entre Ríos (Perisodactyla: Tapiridae). Análisis filogenético preliminar. Jornadas Argentinas de Paleontología de Vertebrados, 21, 2005. Resúmenes. Neuquén, Plaza Huincul.

Ferrero BS and Noriega JI, 2007. A new upper Pleistocene tapir from Argentina: Remarks on the phylogenetics and diversification of neotropical Tapiridae. Journal of Vertebrate Paleontology 27: 504-511.

Forasiepi A, Martinelli A and Blanco J, 2007. *Bestiario Fósil. Mamíferos del Pleistoceno de la Argentina*. Editorial Albatros. Buenos Aires. Pp. 1-190.

Froehlich DJ, 1999. Phylogenetic systematics of basal perissodactyls. Journal of Vertebrate Paleontology 19: 140-159.

'u Y-X, and Li W-H, 1993. Statistical tests of neutrality of mutations. Genetics 133: 693-709.

'u Y-X, 1997. Statistical tests of neutrality against population growth, hitchhiking and background selection. Genetics 147: 915-925.

Gabriel KR and Sokal RR, 1969. A new statistical approach to geographic variation analysis. Systematic Zoology 18: 259-278.

Galvis J, 1980. Un arco de islas Terciario en el Occidente colombiano. Geología Colombiana 11: 7-43.

Garzione CN, Hoke GH, Libarkin JC, Withers S, Macfadden B, Eiler J, Ghosh P, Mulch A, 2008. Rise of the Andes. Science 320: 1245-1380.

Gregorious HR, 1978. The concept of genetic diversity and its formal relationship to heterozygosity and genetic distance. Mathematical Bioscience 41: 253-271.

Haffer J, 1967. Speciation in Colombian forest birds west of the Andes. Amer Mus Novitates 2294: 1-57.

Haffer J. 1969. Speciation in Amazonian forest birds. Science 165: 131-137.

Haffer J, 1982. General aspects of the refuge theory. In: Prance G.T., (Ed.), *Biological diversification in the Tropics*. Columbia University Press, New York, pp. 6-24.

Haffer J, 1997. Alternative models of vertebrate speciation in Amazonia: an overview. Biodivers. Conserv 6: 451-476.

Hamrick JL, Murawski DA and Nason JD, 1993. The influence of seed dispersal mechanisms on the genetic structure of tropical tree populations. Vegetatio 107: 281-297.

Harpending HC, Sherry ST, Rogers AR and Stoneking M, 1993. Genetic structure of ancient human populations. Current Antrophology 34: 483-496.

Harpending HC, 1994. Signature and ancient population growth in a low-resolution mitochondrial DNA mismatch distribution. Human Biology 66:591-600.

Hellberg ME, 1994. Relationships between inferred levels of gene flow and geographic distance in a philopatric coral, *Balanophyllia elegans*. Evolution 48: 1829-1854.

Hershkovitz P, 1954. Mammals of Northern Colombia. Preliminary Report No. 7: Tapirs (genus *Tapirus*), with a systematic review of American species. Proc US Natl Mus Smith Inst 103: 465-496.

Holanda EC and Cozzuol MA. 2006. New records of *Tapirus* from the late Pleistocene of Southwestern Amazonia, Brazil. Revista Brasileira de Paleontologia 9: 193-200.

Holanda EC, Ribeiro AM, Ferigolo J and Cozzuol MA, 2005. Novos registros de *Tapirus* Brunnich, 1771 (Mammalia, Perissodactyla) para o Cuaternario do Brasil. Congreso Latinoamericano de Paleontología de Vertebrados, 2, 2005. Boletim de Resumos, Rio de Janeiro, UFRJ, p 136.

Hooghiemstra H and Van der Hammen T, 2004. Quaternary ice-age dynamics in the Colombian Andes: developing an understanding of our legacy. Philosophical Transaction of the Royal Society B: Biological Sciences 359: 173-181.

Hudson RR, 2000. A new statistic for detecting genetic differentiation. Genetics 155, 2011-2014.

Hudson RR, Boss DD and Kaplan NL, 1992a. A statistical test for detecting population subdivision. Mol Biol Evol 9: 138-151.

Hudson RR, Slatkin M and Maddison WP, 1992b. Estimations of levels of gene flow from DNA sequence data. Genetics 132: 583-589.

Hulbert RC, 1995. The giant tapir, *Tapirus haysii*, from Leisey Shell Pit 1A and other Florida Irvingtonian localities. Bull Fla Mus Nat Hist 37: 515-551.

Hulbert RC, 1999. Nine million years of *Tapirus* (Mammalia, Perissodactyla) from Florida. Journal of Vertebrate Paleontology 19: 53A.

Hulbert RC and Wallace SC, 2005. Phylogenetic analysis of Late Cenozoic *Tapirus* (Mammalia, Perissodactyla). Journal of Vertebrate Paleontology 25: 72A.

Hulbert RC, Wallace SC, Klippel WE and Parmalee PW, 2009. Cranial morphology and systematics of an extraordinary sample of the late Neogene Dwarf Tapir, *Tapirus polkensis* (Olsen). Journal of Paleontology 83: 238-262.

Hyrooka SS, 2003. As cavernas do Bauxi como detentoras de informacoes do periodo Pleistoceno. Simposio de Geologia do Centro-Oeste, 8, 2003. Cuiabá. Boletim de resumos, Cuiabá, UFMT. Pp. 204-205.

Hoffstetter R, 1986. High Andean mammalian faunas during the Plio-Pleistocene. In *High altitude tropical biogeography*. Vuilleumier, F., Monasterio, M. (Eds.). Oxford University Press, New York. Pp. 218-245.

Holbrook LT, 1999. The phylogeny and classification of tapiromorph perissodactyls (Mammalia). Cladistics 15: 331-350.

Isaaks EH and Srivastava RM, 1989. *An introduction to applied geostatistics*. Oxford University Press, New York. Pp 1-561.

Lavergne A, Ruiz-García M, Catzeflis F, Lacote S, Contamin H, Mercereau-Puijalon O, Lacaste A and De Thoisy B, 2010. Taxonomy and phylogeny of squirrel monkey (genus *Saimiri*) using cytochrome b genetic analysis. Am J Primatol. 72, 242-253.

MacFadden BJ, 1992. *Fossil horses: Systematics, paleobiology, and evolution of the family Equidae*. Cambridge University Press, New York.

Manly BFJ, 1997. *Randomization, Bootstrap and Monte Carlo methods in Biology*. Chapman & Hall, London

Mantel NA, 1967. The detection of disease clustering and a generalized regression approach. Cancer Research 27: 209-220.

Marshall LG, Berta A, Hoffstetter R, Pascual R, Reig OA, Bombin M and Mones A, 1984. Mammals and stratigraphy: geochronology of the continental mammal-bearing Quaternary of South America. Palaeovertebrata, Mémoire Extraordinaire, Pp. 1-76.

Matula DW and Sokal RR. 1980. Properties of Gabriel graphs relevant to geographic variation research and the clustering of points in the plane. Geographic Analysis 12: 205-222.

McKenna MC and Bell SK, 1997. *Classification of Mammals-Above the species level*. Columbia University Press, New York. Pp. 631.

Metais G, Soe AN and Ducrocq S, 2006. A new basal tapiromorph (Perissodactyla, Mammalia) from the middle Eocene of Myanmar. Geobis 39: 513-519.

Moran PAP, 1950. Notes on continuous stochastic phenomena. Biometrika 37: 17-23.

Morral N, Bertrantpetit J, Estivill X, 1994. The origin of the major cystic fibrosis mutation (delta F508) in European populations. Nat Genetics 7: 169-175.

Nabel PE, Cione A and Tonni EP, 2000. Environmental changes in the Pampean area of Argentina at the Matuyama-Brumbes (C1r-C1n) Chrons boundary. Palaeogeography, Palaeoclimatology, Palaeoecology 162: 403-412.

Nei M, 1972. Genetic distance between populations. American Naturalist 106: 283-292.

Nei M and Kumar S, 2000. *Molecular Evolution and Phylogenetics*. Oxford University Press, New York. pp. 333.

Nei M, Maruyama T and Chakraborty R, 1975. The bottleneck effect and genetic variability in populations. Evolution 29: 1-10.

Noriega J, Carlini AA, Brandoni O, Ferrero BS, Vassalo C and Cettour de Soto S, 2004. Mamíferos del Cuaternario de la cuenca del río Uruguay, Departamento de Concordia, Entre Ríos, Argentina. Reu. An. de Comunic. APA, Resúmenes, Diamante, APA. Pp. 21-22.

Norman JE and Ashley MV, 2000. Phylogenetics of Perissodactyla and tests of the molecular clock. Journal of Molecular Evolution 50: 11-21.

Norton JE and Ashley MV, 2004a. Genetic variability and population structure among wild Baird's tapirs. Animal Conservation 7: 211-220.

Norton JE and Ashley MV, 2004b. Genetic variability and population differentiation in captive Baird's tapirs (Tapirus bairdii). Zoo Biology 23: 521-531.

Oliveira EV, 1992. Mamíferos Fósseis do Quaternário do Estado do Rio Grande do Sul, Brasil. MSc thesis Dissertation. Programa de Pós-Graduacao em Geociencias, Universidade Federal do Rio Grande do Sul, Porto Alegre. Pp. 1-101.

Oden N, 1984. Assessing the significance of a spatial correlogram. Geographical Analysis 16: 1-16.

Orme DL, Davies RG, Burgués M, Eigenbrod F, Pickup N, Olson VA, Webster AJ, Ding TS, Rasmussen PC, Ridgely RS, Stattersfield AJ, Bennett PM, Blackburn TM, Gaston KJ and Owens IPF, 2005. Global hotspots of species richness are not congruent with endemism or threat. Nature 436: 1016-1019.

Patterson B and Pascual R, 1968. The fossil mammal fauna of South America. The Quarterly Review of Biology 43: 409-451.

Patton JL, da Silva MNF and Malcolm JR, 2000. Mammals of the Rio Jurua and the evolutionary and ecological diversification of Amazonia. Bull Am Mus Nat Hist 244: 1-306.

Paula Couto C, 1980. Fossil mammals of the Pleistocene of Jacupiranga State of Sao Paulo, Brazil. Anais da Academia Brasileira de Ciencias 52: 135-142.

Porpino KO and Santos MFCF dos, 2003. Novos registros de Artiodactyla e Perissodactyla para o Laredo da Escada, Baraúna/RN. Congresso Brasileiro de Paleontología, 18, 2003. Boletim de Resumos, Brasilia, UnB, p. 226.

Pons O and Petit RJ, 1995. Estimation, variance and optimal sampling of gene diversity. I. Haploid locus. Theoretical Applied Genetics 90: 462-470.

Radinsky LB, 1963. Origin and early evolution of North American Tapiroidea. Peabody Museum of Natural History 17: 1-103.

Radinsky LB, 1965. Evolution of the tapiroid skeleton from Heptodon to Tapirus. Bull Mus Comp Zool 134: 69-106.

Radinsky LB, 1967. A review of the rhinocerotoid family Hydracodontidae (Perissodactyla). Bull Am Mus Nat Hist 136: 1-46.

Radinsky LB, 1969. The early evolution of the Perissodactyla. Evolution 23: 308-328.

Rambaut A and Drummond AJ, 2007. Tracer v1.4. Available from http://beast.bio.ed.ac.uk/Tracer.

Ramos-Onsins SE and Rozas J, 2002. Statistical properties of new neutrality tests against population growth. Mol Biol Evol 19: 2092-2100.

Rancy A, 1981. Mamíferos fósseis do Cenozóico do Alto Juruá-Acre. MSc Dissertation. Programa de Pós-Graduacao em Geociencias. Universidade Federal do Rio Grande do Sul. Pp. 122.

Ripley BD, 1981. *Spatial statistics.* John Wiley and Sons, New York.

Rogers AR and Harpending HC, 1992. Population growth makes waves in the distribution of pairwise genetic differences. Mol Biol Evol 9: 552-569.

Rogers AR, Fraley AE, Bamshad MJ, Watkins WS and Jorde LB, 1996. Mitochondrial mismatch analysis is insensitive to the mutational process. Mol Biol Evol 13: 895-902.

Rolim JL, 1974. Paleontologia e estratigrafia do Pleistoceno continental do Nordeste Brasileiro "Formacao Cacimbas". MSc thesis Dissertation. Programa de Pós-Graduacao em Geociencias, Universidade Federal do Rio Grande do Sul, Porto Alegre.

Ruiz-Garcia M, 1998. Genetic structure of different populations of domestic cat in Spain, Italy, and Argentina at a micro-geographic level. Acta Theriologica 43: 39-66.

Ruiz-Garcia M, 1999. Genetic structure of different cat populations in Europe and South America at a microgeographic level: Importance of the choice of an adequate sampling level in the accuracy of population genetics interpretations. Genetics and Molecular Biology 22: 493-505.

Ruiz-García M, 2003. Molecular population genetic analysis of the spectacled bear (*Tremarctos ornatus*) in the Northern Andean Area. Hereditas 138: 81-93.

Ruiz-García M, 2007. Genética de Poblaciones: Teoría y aplicación a la conservación de mamíferos neotropicales (Oso andino y delfín rosado). Bol Real Soc Esp Hist Nat 102 (1-4): 99-126.

Ruiz-García M, 2010a. Micro-geographical genetic structure of *Inia geoffrensis* in the Napo-Curaray River basin by means of Chesser's models. In *Biology, Evolution, and Conservation of River Dolphins Within South America and Asia.* Ruiz-García, M., Shostell, J. (Eds) Nova Science Publishers., Inc.. New York. Pp. 131-160.

Ruiz-García M, 2010b. Changes in the demographic trends of pink river dolphins (*Inia*) at the micro-geographical level in Peruvian and Bolivian rivers and within the Upper Amazon: Microsatellites and mtDNA analyses and insights into *Inia's* origin. In *Biology, Evolution, and Conservation of River Dolphins Within South America and Asia.* Ruiz-García, M., Shostell, J. (Eds.) Nova Science Publishers., Inc.. New York. Pp. 161-192.

Ruiz-García M, 2012. The genetic demography history and phylogeography of the Andean bear (*Tremarctos ornatus*) by means of microsatellites and mtDNA markers. In *Molecular Population Genetics, Phylogenetics, Evolutionary Biology and Conservation of the Neotropical Carnivores.* Ruiz-García, M., Shostell, J. (Eds.). Nova Science Publishers. New York.

Ruiz-Garcia M and Jordana J, 1997. Spatial genetic structure of the "Gos d' Atura" dog breed in Catalonia (Spain). Brazilian Journal of Genetics 225-236.

Ruiz-Garcia M and Jordana J, 2000. Spatial structure and gene flow from biochemical markers in the "Pyrenean Brown" breed, a rare cattle race in Catalonia (Spain). Biochemical Genetics 38: 341-352.

Ruiz-García M, Orozco-terWengel P, Payán E and Castellanos A, 2003. Genética de Poblaciones molecular aplicada al estudio de dos grandes carnívoros (*Tremarctos ornatus* – Oso andino, *Panthera onca*- jaguar): lecciones de conservación. Bol Real Soc Esp Hist Nat 98: (1-4): 135-158.

.uiz-García M, Orozco-terWengel P, Castellanos A and Arias L, 2005. Microsatellite analysis of the spectacled bear (*Tremarctos ornatus*) across its range distribution. Genes and Genetics Systems 80: 57-69.

.uiz-García M, Payán E, Murillo A and Alvarez D, 2006. DNA Microsatellite characterization of the Jaguar (*Panthera onca*) in Colombia. Genes and Genetics Systems 81: 115-127.

.uiz-García M, Randi E, Martínez-Aguero M and Alvarez D, 2007. Relaciones filogenéticas entre géneros de ciervos neotropicales (Artiodactyla, Cervidae) mediante secuenciación de ADN mitocondrial y marcadores microsatelitales. Revista de Biología Tropical. International Journal of Tropical Conservation and Biology. 55: 723-741.

Ruiz-García M, Caballero S, Martínez-Aguero M and Shostell J, 2008. Molecular differentiation among *Inia geoffrensis* and *Inia boliviensis* (Iniidae, Cetacea) by means of nuclear intron sequences. In *Population Genetics Research Progress*. Koven, V.P (Ed.) Nova Science Publisher, Inc. New York. Pp 177-223.

.uiz-García M, Pacheco L and Alvarez D, 2009a. Caracterización genética del puma andino boliviano (*Puma concolor*) en el Parque Nacional Sajama y relaciones con otras poblaciones de pumas del nor-occidente de Sudamérica. Revista Chilena de Historia Natural 82: 97-117.

Ruiz-García M, Martínez-Aguero M, Alvarez D and Goodman S, 2009b. Variabilidad genética en géneros de Ciervos neotropicales (Mammalia: Cervidae) según loci microsatélitales. Revista de Biología Tropical. International Journal of Tropical Conservation and Biology. 57: 879-904.

Ruiz-García M and Pinedo-Castro M, 2010. Molecular Systematics and Phylogeography of the genus *Lagothrix* (Atelidae, Primates) by means of mitochondrial COII gene. Folia Primatol 81: 109-128.

Ruiz-García M, Castillo MI, Vásquez C, Rodríguez K, Pinedo M, Shostell J and Leguizamon N, 2010. Molecular Phylogenetics and Phylogeography of the White-fronted capuchin (*Cebus albifrons*; Cebidae, Primates) by means of mtCOII gene sequences. Mol Phylogenet Evol 57: 1049-1061.

Ruiz-García M, Vásquez C, Camargo E, Leguizamon N, Gálvez H, Vallejo A, Pinedo-Castro M, Castellanos-Mora LF, Shostell J and Alvarez D, 2011a. The molecular phylogeny of the *Aotus* genus (Cebidae, Primates). Int J Primatol. 32: 1218-1241.

Ruiz-García M, Castillo MI, Ledezma A, Leguizamon N, Sánchez R, Chinchilla M and Gutierrez-Espeleta G, 2011b. Molecular Systematics and Phylogeography of *Cebus capucinus* (Cebidae, Primates) in Colombia and Costa Rica by means of mitochondrial COII gene. Am J Primatol. 73: 1-15.

Ruiz-García M, Lichilín N, Gutierrez-Espeleta G, Escobar-Armel P, Castillo MI, Thoisy B and Wallace R, 2012a. Molecular phylogeny of the *Ateles* genus by means of the mitochondrial COII gene and nuclear DNA microsatellites. Mol Phylogenet Evol (submitted).

Ruiz-García M, Vásquez C, Sandoval S, Thoisy B and Kaston F, 2012b. Genetic structure, phylogeography and demographic evolution of the Lowland tapir (*Tapirus terrestris*) with special emphasis in upper Amazon and Colombia. (to be submitted).

Rusconi C, 1928. Nueva especie fósil de tapir de la Argentina *Tapirus australis* n. sp. Con una nota sobre *Tapirus tarijensis*, Ameg. Imp ML Rañó. Pp. 3-15.

Saillard J, Forster P, Lynnerup N, Bandelt H-J and Norby S, 2000. mtDNA variation among Greenland Eskimos: the edge of the Beringian expansion. American Journal o Hum. Genet 67: 718-726.

Salles LO, Cartelle C, Guedes PG, Boggiani PC, Janoo A and Russo CAM. 2006. Quaternary mammals from Serra da Bodoquena, Mato Grosso do Sul, Brazil. Boletim do Museu Nacional-Série Zoologia 521: 1-12.

Sambrook J, Fritsch EF and Maniatis T, 1989. Molecular cloning: A laboratory manual. 2nd ed. New York: Cold Spring Harbor Laboratory Press.

Savage RJG and Long MR, 1986. Evolución de los Mamíferos: Una guía ilustrada. Ediciones Akal, Torrejón de Ardoz, Madrid. Pp. 258.

Sedano RE, Burns KJ, 2010, Are the Northern Andes a species puma for Neotropical birds . Phylogenetics and biogeography of a clade of Neotropical tanagers (Aves Thraupini). Journal of Biogeography 37: 325-343.

Sedor FA, Born PA and Santos FMS, 2005. Fósseis pleistocénicos de Scelidodon (Mylodontidae) e Tapirus (Tapiridae) em cavernas paranaenses (PR, sul do Brasil) Acta Biológica Paraense 33: 121-128.

Simonsen KL, Churchill GA and Aquadro CF, 1995. Properties of statistical tests o neutrality for DNA polymorphism data. Genetics 141: 413-429.

Simpson GG, 1950. History of the fauna of Latin America. Amer Scien 38: 361-389.

Simpson GG, 1965. The geography of evolution. Collected essays. Philadelphia and New York, Chilton Books.

Simpson GG, 1980. Splendid isolation, the curious history of South American mammals Yale University Press, New Haven, 266pp.

Smouse PE, Long JC and Sokal RR, 1986. Multiple regression and correlation extension o the Mantel test of matrix corresponde. Syst Zool 35: 627-632.

Sokal RR, Harding RM and Oden NL, 1989. Spatial patterns of human gene frequencies ir Europe. American Journal of Physical Anthropology 80: 267-294.

Sokal RR and Jacquez GM, 1991. Testing inferences about microevolutionary processes by means of spatial autocorrelation analysis. Evolution 45: 155-172.

Sokal RR and Oden NL, 1978a. Spatial autocorrelation in Biology. 1. Methodology Biological Journal of Linnean Society 10: 199-228.

Sokal RR and Oden NL, 1978b. Spatial autocorrelation in Biology. 2. Some biological implications and four applications of evolutionary and ecological interest Biological Journal of Linnean Society 10: 229-249.

Sokal RR, Oden NL and Barker JSF, 1987. Spatial structure in *Drosophila buzzatii* populations simple and directional spatial autocorrelation. American naturalist 129: 122-142.

Sokal R and Rohlf FJ, 1995. *Biometry*. 3rd edition. W.H. Freeman and Co., New York

Sokal RR, Smouse PE and Neel JV, 1986. The genetic structure of a tribal population, the Yanomama Indians. Genetics. XV. Patterns inferred by autocorrelation analysis Genetics 114: 259-287.

Sokal RR and Wartenberg DE, 1983. A test of spatial autocorrelation using and isolation by distance model. Genetics 105: 219-237.

Soliani E. 1973. Geologia da regiao de Santa Vitória do Palmar, Rio Grande do Sul, e a posicao estratigrafica dos fósseis de mamíferos Pleistocenicos. MSc thesis

Dissertation. Programa de Pós-Graduacao em Geociencias, Universidades Federal do Rio Grande do Sul, Porto Alegere. Pp. 1-88.

ouza-Cunha FL. 1959. *Mamíferos do Pleistoceno do Rio Grande do Sul. I. Ungulados.* DNPM, DGM, Rio de Janeiro. Pp. 47.

passov N and Ginsburg L, 1999. *Tapirus balkanicus* nov. sp., nouveau tapir (Perissodactyla, Mammalia) du Turolien de Bulgarie. Annales de Paléontologie 85: 265-276.

phuler JN, 1972. Genetic, linguistic, and geographical distances in native North America. In *The assessment of population affinities in man.* Wiener, J. S., Huizinga, J (Eds.). Oxford University Press, Oxford.

treiff R, Labbe T, Bacilieri R, Steinkellner H, Glossl J and Kremer A, 1998. Within-population genetic structure in *Quercus robur* L and *Quercus petraea* (Matt.) Liebl. assessed with isoenzymes and microsatellites. Molecular Ecology 7: 317-328.

ajima F, 1989. Statistical method for testing the neutral mutation hypothesis by DNA polymorphism. Genetics 123: 585-595.

amura K, 1992 Estimation of the number of nucleotide substitutions when there are strong transition-transversion and G+C content biases. Mol Biol Evol 9: 678-687.

apia A, Arévalo M, Sánchez ME, Witte T, Llumipanta W and Paz-y-Miño C, 2006. Using PCR-SSCP as tool to detect polymorphism in tapirs. Tapir Conservation- The newsletter of the IUCN/SSC Tapir specialist group- 15: 37-39.

hoysi B, Goncalves da Silva A, Ruiz-García M, Tapia A, Ramirez O, Arana M, Quse V, Paz-y-Miño C, Tobler T, Pedraza C, Lavergne A. 2010. Population history, phylogeography, and conservation genetics of the last Neotropical mega-herbivore, the Lowland tapir *(Tapirus terrestris).* BMC Evolutionary Biology 10: 278-295.

irira D, 2008. *Mamíferos de los bosques húmedos del noroccidente de Ecuador.* Ediciones Murciélago blanco y Proyecto PRIMENET. Publicación Especial sobre los Mamíferos de Ecuador 7. Quito. 352 pp.

onni EP, 1992. *Tapirus* Brisson, 1762 (Mammalia, Perissodactyla) en el Lujanense (Pleistoceno Superior-Holoceno Inferior) de la Provincia de Entre Ríos, Republica Argentina. Ameghiniana 29: 3-8.

onni EP and Pasquali RC, 2005. *Mamíferos Fósiles. Cuando en las Pampas vivían los gigantes.* Editorial Científica Universitaria. Córdoba.

ougard C, Delefosse T, Hänni C and Montgelard C, 2001. Phylogenetics relationships of the five extant Rhinoceros species (Rhinocerotidae, Perissodactyla) based on mitochondrial cytochrome b and 12S rRNA genes. Molecular Phylogentic and Evolution 19: 34-44.

rexler JC, 1988. Hierarchical organization of genetic variation in the sailfin molly, *Poecilia latipinna* (Pisces: Poeciliidae). Evolution 42: 1006-1017.

subamoto T, Egi N, Takai M, Sein C and Hamhung M, 2005. Middle Eocene ungulate mammals from Myanmar: a review with description of new specimens. Acta Palaeontologica Polonica 50: 139-150.

Jbilla M, 1983. Sobre la presencia de tapires fósiles en el Uruguay (Mammalia, Perissodactyla, Tapiridae). Revista de la Facultad de Humanidades y Ciencias 1: 85-104.

Jbilla M, 1996. Paleozoologia del Cuaternario continental de la cuenca norte del Uruguay: biogeografía, cronología y aspectos climático-ambientales. Programa de desarrollo en Ciencias Básicas, Universidad de La República, Uruguay. PhD Thesis Dissertation. Pp. 232.

Ubilla M and Rinderknecht A, 2006. Un Nuevo registro de tapir (Mammalia: Tapiridae) para el Pleistoceno del norte de Uruguay (formación Sopas). Jornadas Argentinas de paleontología de Vertebrados, 22, 2006, San Juan. Resúmenes, San Juan. Pp. 33-34.

Upton G and Fingleton B, 1985. *Spatial data analysis by example*. Vol 1: Point pattern and quantitative data. John Wiley and Sons, Chichester.

Van der Hammen T, 1961. The Quaternary climatic changes of northern South America Ann New York Acad Sciences 95: 676-683.

Van der Hammen T, 1980. Glaciales y glaciaciones en el Cuaternario de Colombia palaeoecología y estratigrafía. Primer Seminario sobre El Cuaternario en Colombia Bogotá DC., Pp. 44-45.

Van der Hammen T, 1992. *Historia, Ecología y Vegetación*. Corporación colombiana para la Amazonía-Araracuara- Bogotá, Colombia. Pp. 411.

Van Geel H and Van der Hammen T, 1973. Upper Quaternary vegetational and climatic sequence of the Fúquene area (Eastern Cordillera, Colombia). Palaeogeography Palaeoclimatology Palaeoecology 14: 9-92.

Vendramin GG, Degen B, Petit RJ, Anzidei M, Madaghiele A and Ziegenhagen B, 1999. High level of variation at *Abies alba* chloropast microsatellite loci in Europe. Molecular Ecology 8: 1117-1126.

Walsh PS, Metzger DA and Higuchi R, 1991. Chelex 100 as a medium for simple extraction of DNA for PCR-based typing from forensic material. BioTechniques 10, 506-513.

Webb DS, 2006. The great American biotic interchange: Patterns and processes. Annals of the Missouri Botanical Garden 93: 245-257.

Whitmore TC and Prance GT, 1987. *Biogeography and Quaternary History in Tropical America*. In: Oxford Monographs in Biogeography, vol. 3. Oxford University Press, Oxford.

Winge H, 1906. Jordfundne og nulevende Hoydyr (Ungulata) fra Lagoa Santa, Minas Gerais. E. Mus Lundii 3: 1-239.

3

Ecological Factors that Influence Genetic Structure in *Campylobacter coli* and *Campylobacter jejuni*

Helen M. L. Wimalarathna[1] and Samuel K. Sheppard[1,2]
[1]*The University of Oxford, Department of Zoology,*
[2]*The University of Swansea, College of Medicine*
United Kingdom

1. Introduction

Campylobacter is the leading cause of human bacterial gastroenteritis worldwide (Friedman *et al.* 2000). Campylobacteriosis, caused principally by the organisms *C. jejuni* and *C. coli*, is characterized by severe diarrhoea, usually accompanied by fever, abdominal pain, nausea and malaise (Allos 2001). Campylobacter infection accounts for an estimated 2.5 million cases of gastro-intestinal disease in the United States and 1.3 million cases in the United Kingdom each year (Kessel *et al.* 2001), and the estimated economic burden of Campylobacter infection is $8 billion in the US and £500 million in the UK. Though pathogenic in humans, these *Campylobacter* species are wide-spread commensals in the digestive tracts of many wild and domesticated animals. Because of its public health significance, considerable effort has gone into understanding how this common organism is transmitted from these reservoir hosts to humans through contaminated meat, poultry, water, milk and contact with animals (Niemann *et al.* 2003).

1.1 Molecular typing

Molecular typing of pathogenic bacteria has enhanced many epidemiological studies, including the identification of food-borne outbreaks of infection due to *E. coli* O157: H7 (Bender *et al.* 1997), *Salmonella enteritica* (Bender *et al.* 2001) and *Listeria monocytogenes* (Olsen *et al.* 2005) and early identification of an outbreak source can enable effective disease containment (Olsen *et al.* 2005) (Rangel *et al.* 2005). However, in many species of bacteria it is impossible to predict the lineage from which a pathogenic phenotype will arise. For example, in *Bacillus cereus*, pathogenicity is associated with mobile elements which mediate spore-formation and toxin production (Raymond *et al.* 2010). These elements can be acquired by distantly related lineages and it is, therefore, difficult to predict the likelihood that a strain will be pathogenic from genotypes derived from non-plasmid DNA alone.

In species such as *Campylobacter*, particular genetically related groups often display similar disease associated phenotypes. The consistency of *Campylobacter* genotypes within sub-populations, and the variation between sub-populations can be exploited in order to determine the source of human infection by comparing clinical isolate genotypes data with large reference sets isolated from known host-species.

Methods such as PFGE and serotyping have shown that far from being monomorphic pathogenic clones like *Yersinia pestis* or *Mycobacterium lepris* (Achtman 2008), *Campylobacter* populations are highly structured with complex associations among lineages at different levels of relatedness. The DNA sequence-based typing method of Multi Locus Sequence Typing (MLST) has provided considerable insight into population structure in recombining organisms such as *Campylobacter*. MLST is an unambiguous high-resolution genotyping method, exploiting genetic variation in fragments of seven separate housekeeping genes. Each locus is approximately 500bp in length, with a defined start and end point. Each unique sequence at a given locus is assigned an allele number, and a Sequence Type (ST) is identified by a unique series of seven numbers, referring to the specific alleles present at each locus. Related STs can be grouped into Clonal complexes, in which sequences are identical at four or more loci (Maiden *et al.* 1998).

This high degree of genetic structuring is the result of a complex interplay of mutation, which leads to the gradual divergence of clonally related lineages, and horizontal gene transfer (HGT), that can lead to the replacement of homologous DNA with sequence from another lineage or in extreme cases the introduction of new genes. While *Campylobacter* can be highly recombinogenic (Wilson *et al.* 2009), mutation and HGT have not been sufficient to erase the clonal signal of descent from the genomes and in the following sections we will investigate some of the ways in which this high degree of genetic structuring can tell us about the biology of this organism and how this relates to disease.

2. Disease-associated lineages

Analysis of the genotypes of *Campylobacter* isolated from human disease cases has shown that the vast majority of campylobacteriosis cases are caused by *C. jejuni* and *C. coli* lineages also found in other potential disease reservoirs, particularly chickens and cattle (Figure 1). Most human disease strains also occur as commensal organisms in domesticated animals, and clinical isolates are a non-random subset of these lineages. This is particularly marked in the case of *C. coli*, in which the average diversity per locus is 13 alleles in disease cases compared with 55 alleles in the general *C. coli* population.

The apparent absence of asymptomatic carriage of *C. jejuni* and *C. coli* among individuals in the UK suggests that humans may not be a natural host for these organisms in high income countries, and have undergone a relatively short history of co-evolution. For this reason, an appreciation of phylogenetic relationships, together with an examination of the ecology of *Campylobacter* can enhance understanding of the origin and causes of human campylobacteriosis and this forms the basis for much of the recent work to explain the epidemiology of these organisms (de Haan *et al.* 2010; Hastings *et al.* 2011; Jorgensen *et al.* 2011).(Kittl *et al.* 2011; Lang *et al.* 2010; Magnusson *et al.* 2011; Mullner *et al.* 2010; Sheppard *et al.* 2011a; Sproston *et al.* 2011; Sproston *et al.* 2010; Thakur & Gebreyes 2010).

3. Contrasting population structure of *Campylobacter jejuni* and *Campylobacter coli*

A neighbour joining tree (Figure 2) shows that *C. jejuni* and *C. coli* display markedly different population structures. *C. jejuni* populations are highly structured into clonal omplexes (Figure 3), clusters of related lineages that share alleles at four or more MLST ɔci. When each locus is considered separately, there is evidence of considerable ecombination within *C. jejuni*, with alleles from disparate locations around the tree ippearing within the same STs. In contrast, *C. coli* displays far greater genetic diversity, with hree deep-branching clades, of which clade 1 contains the vast majority of lineages lescribed to date. The ST-828 clonal complex, part of clade 1, accounted for around 70% of he 2289 *C. coli* isolates submitted to the PubMLST database (http://pubmlst.org/ ampylobacter) before September 5th 2011, with most of the remainder sharing alleles with hese, and therefore also being related. The second most common clonal complex (ST-1150 ɔmplex), also from clade 1, accounts for only 2% of isolates.

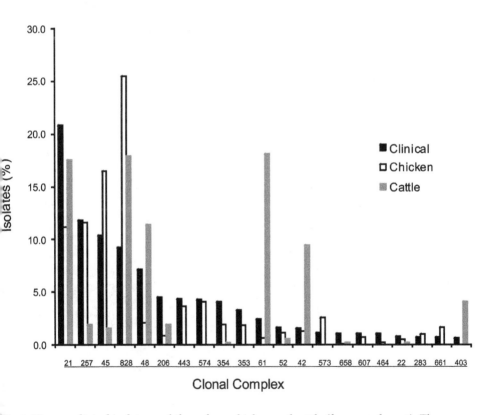

Fig. 1. Human clinical isolates and those from chicken and cattle (faeces and meat). The relative abundance of clonal complexes (responsible for >1% of total UK disease) of isolates from human *C. jejuni* and *C. coli* infections and those from published chicken and cattle isolate collections (Sheppard *et al.* 2010a; Sheppard *et al.* 2009b; Sheppard *et al.* 2010b). All of the 21 most common disease causing clonal complexes are also found in cattle, chickens or both.

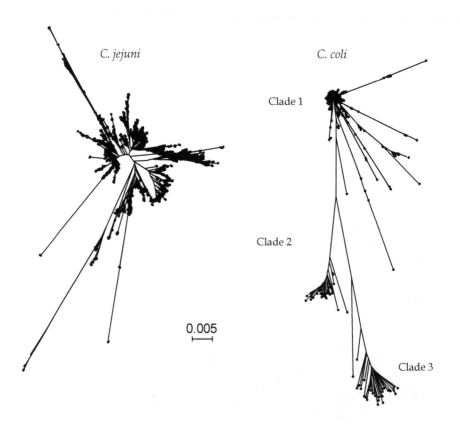

Fig. 2. The genetic relatedness of 1341 *C. jejuni* and *C. coli* genotypes based on concatenated MLST alleles (3309bp) from published studies (Sheppard *et al.* 2010a; Sheppard *et al.* 2010b). Contrasting tree topologies are visible on the neighbour-joining trees with three deep branching clades present among *C. coli* genotypes.

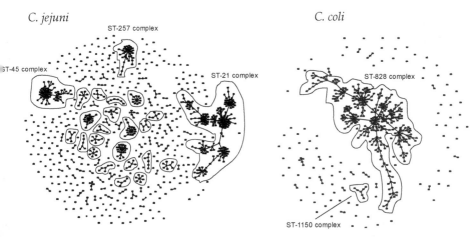

'ig. 3. Comparison of the population structure of *C. jejuni* and *C. coli*. Different 7-locus ;enotypes are represented by points on a goeBURST diagram; strains differing at a single ocus are joined by a lines that infer linkage by decent. Cluster size distribution is different or the two species with many more clonal complexes found among *C. jejuni* genotypes that vithin *C. coli* where most of the typed strains belong to the ST-828 complex.

3.1 Trefoil structure in *Campylobacter coli*

The emergence and maintenance of the 3-clade structure in *C. coli* implies that three distinct »acterial gene pools exist, and that although recombination is evident within *C. coli* clades, here are or have been barriers to recombination between clades. Recombinational barriers ·an be considered in three broad categories: (i) adaptive, implying selection against hybrids; ii) mechanistic, imposed by homology dependence of recombination or other factors »romoting DNA specificity; (iii) ecological, a consequence of physical separation in different •cological niches (Sheppard *et al.* 2010b). In order to consider the relative importance of each ype of barrier in the evolution of a clade structure in *C. coli* it may be useful to look at both *C. coli* and *C. jejuni* in context.

Campylobacter jejuni and *C. coli* are approximately 12% divergent at the nucleotide level and ire considered distinct microbial species, however there is strong evidence for a degree of ıybridisation between the species through a process of horizontal gene transfer (HGT) (Sheppard *et al.* 2008) (Sheppard *et al.* 2011b). Statistical model-based approaches have been ısed to investigate the sharing of both whole alleles and recombined elements or 'mosaic ılleles' between *C. jejuni* and *C. coli*. While *C. coli* clade 1 remains distinct from clades 2 & 3, here is evidence of gene flow between *C. jejuni* and *C. coli* clade 1. Analysis of 1738 alleles 'rom a total of 2953 Sequence Types identified 31 mosaic alleles, of which 25 had demonstrably been acquired by *C. coli* from *C. jejuni*, and the remaining 6 had originated in *C. coli* and been acquired by *C. jejuni*. With the exception of a single mosaic allele, having »riginated in *C. coli* clade 3 and being acquired by *C. jejuni*, all genetic exchange events dentified involved *C. coli* clade 1 as either donor or recipient.

The existence of hybrids and the maintenance of alleles of *C. jejuni* origin within the *C. coli* gene pool demonstrates that mechanistic barriers are not preventing interspecies gene flow. Furthermore, the resultant hybrid lineages are not sufficiently maladapted to prevent their proliferation (adaptive barrier). Ecological barriers to recombination are therefore likely to have been important in generating and maintaining the observed population structure in *C. coli* and *C. jejuni* species, clades and clonal complexes.

4. Ecology and host association

There is evidence of association between clusters of related genotypes and the source or host from which the bacteria were isolated. At the species level, *C. jejuni* and *C. coli* have subtly different host ranges. Both species are found in a wide range of wild and farm animals but *C. jejuni* dominate numerically in most sampled wild bird species (Colles *et al.* 2008a, Sheppard *et al.* 2011a) as well as chickens and cattle (Sheppard *et al.* 2009a). *C. coli* (clade 1) are also common in chicken and cattle, usually constituting around 10% of the *Campylobacter* population in these host animals (90% *C. jejuni*); but are more abundant than *C. jejuni* in pigs (Miller *et al.* 2006). Within *C. coli*, isolates belonging to clades 2 & 3 are far less common and are usually isolated from environmental sources where they may be associated with waterfowl.

The host-genotype relationship goes further than this. In *C. coli* and *C. jejuni* there is a strong association between specific clonal complexes (mainly *C. jejuni*), STs, and alleles and host species (McCarthy *et al.* 2007; Miller *et al.* 2006; Sheppard *et al.* 2010a). This association is stronger than spatial or temporal signals and statistical assignment analyses consistently correctly grouped isolates from a range of host animal sources regardless of geographical source (Sheppard *et al.* 2010a). For example, a population of *C. jejuni* isolates from UK chickens is strikingly similar to a population of *C. jejuni* isolates from chickens in the US, mainland Europe or Senegal. The equivalent is true of cattle, pigs and turkeys (Sheppard *et al.* 2010a). This host allelic signature between diverse lineages inhabiting the same ecological niche creates a pool of alleles common to a given source (McCarthy *et al.* 2007) and this signal of host association has been widely used to assign the origin reservoir of clinical isolates (Mullner *et al.* 2009; Sheppard *et al.* 2009b; Strachan *et al.* 2009; Wilson *et al.* 2008). All of these studies identify farm (especially chicken) associated isolates as the main source of human infection.

5. Why do farm associated isolates cause disease?

There are two possible explanations for the strong correlation between genotypes that cause human disease, and those that are associated with farm animals, especially chickens and ruminants. First, this could be the result of differential exposure. By definition, humans are more frequently exposed to domesticated food animals than to wild reservoirs of infection. The main risk factors for human campylobacteriosis include handling and consumption of raw or under-cooked poultry (Kapperud *et al.* 1992) (Friedman *et al.* 2004); handling and consumption of barbequed meat (Studahl& Andersson 2000); contact with farm animals (Friedman *et al.* 2004) and consumption of unpasteurised milk (Niemann *et al.* 2003). These risk factors are all common behaviours which present opportunities for exposure to domesticated animals and animal products, whilst exposure to wild and environmental sources of *Campylobacter* may be less common. It is therefore possible that all *Campylobacter* strains are equally infective and the dominance of farm associated genotypes in human disease is simply reflective of greater exposure to these strains.

Alternatively, it is possible that certain strains are more likely to cause acute infection than others. If it were the case that agricultural strains were more pathogenic to humans then they would be over represented in surveys of reported clinical cases. While this may be a less likely explanation than simple differential exposure, some genotypes do appear particularly well adapted to very specific ecological niches and in an evolutionary trade-off their ability to colonise diverse hosts may have been lost. There are numerous examples of host restricted STs among strains found only in specific wild bird species (Waldenstrom *et al.* 2007) (Colles *et al.* 2008b). Genetic isolation could explain this but different colonization capacity could also be important. For example, *C. jejuni* strains (ST 3704) that are routinely found in the gut of bank voles are unable to colonise the chicken gut in laboratory experiments (Williams *et al.* 2010). In a similar experiment, using a European Robin (*Erithacus rubecula*) infection model, *C. jejuni* from song thrushes (*Turdus philomelos*) successfully colonized but *C. jejuni* from human disease did not (Waldenstrom *et al.* 2010).

As already mentioned, *C. jejuni* and *C. coli* have different host ranges and there is evidence that they exhibit colonisation and virulence factors differentially in response to different growth conditions, which may relate to host preferences (Leach *et al.* 1997). For example, there are a wide range of carbon sources that *C. jejuni* utilize more effectively at 42°C rather than the lower temperature of 37°C (Line *et al.* 2010). The average core temperature of a chicken is 42°C, while a pig is 39°C so this could influence the ability of *C. jejuni* to colonize different hosts. Serine dehydratase, encoded by the *sdaA* gene has been demonstrated to be an essential colonisation factor in *C. jejuni*. This gene is also expressed in *C. coli*, but the functionality of the enzyme is highly dependent on temperature. In *C. coli* there is little or no serine dehydratase activity at 42°C, but at the lower temperature of 37°C activity is significantly increased, this could provide a partial explanation for the porcine host association with *C. coli*.

Colonisation and virulence factors in *Campylobacter* are not well understood, but evidence of differential abilities to invade the cells of different hosts points to a possible explanation for the relationship between specific STs and human disease. Explanations based on differential exposure and colonization capacity are not mutually exclusive. It is plausible that those lineages that are found in a niche to which humans are routinely exposed have acquired the necessary colonisation factors to persist in this environment, and opportunistically to infect humans.

6. Dating lineage divergence

Genotyping isolates from various sources can offer insight into the causes of the genetic structuring in *Campylobacter* populations. However, a more comprehensive understanding of the evolution of the genus can be obtained if the time scale for the divergence of lineages can be overlaid upon the tree of genetic relatedness. By cross-referencing estimated dates of divergence within the genus *Campylobacter* with ecological data it is possible to make inferences about the conditions which created the specific barriers which led to speciation, the formation of the lineage structure, and the gene-flow between certain clades and clonal complexes.

The traditional method for dating bacterial evolution is based on the rate of sequence divergence between *Escherichia coli* and *Salmonella typhimurium*, which is assumed to be 1% 16S rRNA divergence per 50 million years (Ochman & Wilson 1987). Applying this method to the *Campylobacter* genus estimates the *C. coli* – *C. jejuni* split to have occurred approximately 10 million years ago, and the divergence of 3 *C. coli* clades about 2.5 million years ago. An alternative dating method, using a molecular clock based on intra-specific diversity in *C. jejuni*, places these splitting events much more recently (Wilson *et al.* 2009). The speciation of *C. coli* and *C. jejuni* has been estimated to have occurred around 6,500 years ago, with *C. coli* clade divergence occurring 1,000-1,700 years ago (Sheppard *et al.* 2010b). While this large disparity between estimates is difficult to explain, there are reasons for favouring the more recent estimates for *Campylobacter* divergence. Methods that provide recent estimates are based on knowledge of genetic variation within the genus *Campylobacter* and not on the divergence of genera (*E. coli* and *S. typhimurium*) only distantly related to *Campylobacter*; additionally there is an increasing number of studies that use similar approaches and infer a more rapid rate of molecular evolution than in traditional models of bacterial evolution (Falush *et al.* 2001; Feng *et al.* 2008; Perez-Losada *et al.* 2007; Wilson *et al.* 2009).

If the diversification leading to the population structure in extant *Campylobacter* populations is placed within the last 6,500 years then it correlates with important changes in human behaviour. For example, the development of agriculture, which began in the middle east about 10,000 years ago and became common in Europe about 5,000-3,00 BC Ammerman & Cavalli-Sforza 1984; McCorriston & Hole 1991; Zvelebil & Dolukhanov 1991) or the establishment of the first cities and the rise of urbanization. Clearly this could have provided novel opportunities for *Campylobacter* to expand into new host species and infect humans in a way that is, to some extent, mirrored in modern society and may have begun to shape the population structure that we observe today.

7. Conclusion

It is evident that the genetic structure that has been described in *C. coli* and *C. jejuni* populations is related to phenotypic factors, such as the animal host from which the isolate was sampled. Furthermore, experimental infections show that genotype is a strong predictor of the host-specific behaviour of a given isolate. Practical applications have effectively exploited this ecology-driven genetic differentiation to attribute the source of human infection but many questions remain about the nature of the forces that result in the highly diverse *Campylobacter* populations. For example, the host association of a particular MLST allele may be influenced by numerous factors including selection for isolates containing particular alleles at loci elsewhere in the genome. As whole genome data become available for large, phenotypically variable isolate collections it will become easier to identify the gene networks that are involved in particular adaptive processes. This has the potential to enhance phylogenetic analysis of *Campylobacter*, and other bacteria, by directly linking the observed population genetic structure and the evolutionary forces that generated it.

!. References

\chtman M (2008) Evolution, population structure, and phylogeography of genetically monomorphic bacterial pathogens. *Annual Review of Microbiology* 62, 53-70.

\llos B (2001) *Campylobacter jejuni* infections: update on emerging issues and trends. *Clinical Infectious Diseases* 2001:32, 1201-1206.

\mmerman AJ, Cavalii-Sforza LL (1984) *The neolithic transition and the genetics of populations in Europe.* Princeton University Press, USA.

\ender JB, Hedberg CW, Besser JM, *et al.* (1997) Surveillance by molecular subtype for *Escherichia coli* O157:H7 infections in Minnesota by molecular subtyping. *New England Journal of Medicine* 337, 388-394.

\ender JB, Hedberg CW, Boxrud DJ, *et al.* (2001) Use of molecular subtyping in surveillance for *Salmonella enterica* serotype typhimurium. *New England Journal of Medicine* 344, 189-195.

\olles FM, Dingle KE, Cody AJ, Maiden MC (2008a) Comparison of *Campylobacter* populations in wild geese with those in starlings and free-range poultry on the same farm. *Applied and Environmental Microbiology* 74, 3583-3590.

\olles FM, Jones TA, McCarthy ND, *et al.* (2008b) *Campylobacter* infection of broiler chickens in a free-range environment. *Environmental Microbiology* 10, 2042-2050.

\e Haan CP, Kivisto RI, Hakkinen M, Corander J, Hanninen ML (2010) Multilocus sequence types of Finnish bovine Campylobacter jejuni isolates and their attribution to human infections. *BMC Microbiol* 10, 200.

\alush D, Kraft C, Taylor NS, *et al.* (2001) Recombination and mutation during long-term gastric colonization by *Helicobacter pylori*: estimates of clock rates, recombination size, and minimal age. *Proceedings of the National Academy of Sciences USA* 98, 15056-15061.

\eng L, Reeves PR, Lan R, *et al.* (2008) A recalibrated molecular clock and independent origins for the cholera pandemic clones. *PLoS ONE* 3, e4053.

\riedman CJ, Neiman J, Wegener HC, Tauxe RV (2000) Epidemiology of *Campylobacter jejuni* infections in the United States and other industrialised nations. In: *Campylobacter* (eds. Nachamkin I, Blaser MJ), pp. 121-138. ASM Press, Washington, D.C.

\riedman CR, Hoekstra RM, Samuel M, *et al.* (2004) Risk factors for sporadic *Campylobacter* infection in the United States: A case-control study in FoodNet sites. *Clinical Infectious Disases* 38 Suppl 3, S285-296.

\astings R, Colles FM, McCarthy ND, Maiden MC, Sheppard SK (2011) Campylobacter genotypes from poultry transportation crates indicate a source of contamination and transmission. *J Appl Microbiol* 110, 266-276.

\orgensen F, Ellis-Iversen J, Rushton S, *et al.* (2011) Influence of season and geography on Campylobacter jejuni and C. coli subtypes in housed broiler flocks reared in Great Britain. *Appl Environ Microbiol* 77, 3741-3748.

\apperud G, Lassen J, Ostroff SM, Aasen S (1992) Clinical features of sporadic Campylobacter infections in Norway. *Scand J Infect Dis* 24, 741-749.

\essel AS, Gillespie IA, O'Brien SJ, *et al.* (2001) General outbreaks of infectious intestinal disease linked with poultry, England and Wales, 1992-1999. *Commun Dis Public Health* 4, 171-177.

Kittl S, Kuhnert P, Hachler H, Korczak BM (2011) Comparison of genotypes and antibiotic resistance of Campylobacter jejuni isolated from humans and slaughtered chickens in Switzerland. *J Appl Microbiol* 110, 513-520.

Lang P, Lefebure T, Wang W, *et al.* (2010) Expanded multilocus sequence typing and comparative genomic hybridization of Campylobacter coli isolates from multiple hosts. *Appl Environ Microbiol* 76, 1913-1925.

Leach S, Harvey P, Wali R (1997) Changes with growth rate in the membrane lipid composition of and amino acid utilization by continuous cultures of Campylobacter jejuni. *J Appl Microbiol* 82, 631-640.

Line JE, Hiett KL, Guard-Bouldin J, Seal BS (2010) Differential carbon source utilization by Campylobacter jejuni 11168 in response to growth temperature variation. *Microbiol Methods* 80, 198-202.

Magnusson SH, Guethmundsdottir S, Reynisson E, *et al.* (2011) Comparison of Campylobacter jejuni isolates from human, food, veterinary and environmental sources in Iceland using PFGE, MLST and fla-SVR sequencing. *J Appl Microbiol.*

Maiden MCJ, Bygraves JA, Feil E, *et al.* (1998) Multilocus sequence typing: a portable approach to the identification of clones within populations of pathogenic microorganisms. *Proceedings of the National Academy of Sciences USA* 95, 3140-3145.

McCarthy ND, Colles FM, Dingle KE, *et al.* (2007) Host-associated genetic import in Campylobacter jejuni. *Emerging Infectious Diseases* 13, 267-272.

McCorriston J, Hole F (1991) The ecology of seasonal stress and the origin of agriculture in the Near East. *American Anthropologist* 93, 46-69.

Miller WG, Englen MD, Kathariou S, *et al.* (2006) Identification of host-associated alleles by multilocus sequence typing of Campylobacter coli strains from food animals. *Microbiology* 152, 245-255.

Mullner P, Collins-Emerson JM, Midwinter AC, *et al.* (2010) Molecular Epidemiology of Campylobacter jejuni in a Geographically Isolated Country with a Uniquely Structured Poultry Industry. *Applied and Environmental Microbiology* 76, 2145-2154.

Mullner P, Jones G, Noble A, *et al.* (2009) Source attribution of food-borne zoonoses in New Zealand: a modified Hald model. *Risk Anal* 29, 970-984.

Niemann J, Engberg J, Molbak K, Wegener HC (2003) A case-control study of risk factors for sporadic Campylobacter infections in Denmark. *Epidemiology and Infection* 130, 353-366.

Ochman H, Wilson AC (1987) Evolution in bacteria: evidence for a universal substitution rate in cellular genomes. *J Mol Evol* 26, 74-86.

Olsen SJ, Patrick M, Hunter SB, *et al.* (2005) Multistate outbreak of *Listeria monocytogenes* infection linked to delicatessen turkey meat. *Clinical Infectious Disases* 40, 962-967.

Perez-Losada M, Crandall KA, Zenilman J, Viscidi RP (2007) Temporal trends in gonococcal population genetics in a high prevalence urban community. *Infect Genet Evol* 7, 271-278.

Rangel JM, Sparling PH, Crowe C, Griffin PM, Swerdlow DL (2005) Epidemiology of *Escherichia coli* O157:H7 outbreaks, United States, 1982-2002. *Emerging Infectious Diseases* 11, 603-609.

Raymond B, Wyres KL, Sheppard SK, Ellis RJ, Bonsall MB (2010) Environmental factors determining the epidemiology and population genetic structure of the Bacillus cereus group in the field. *PLoS Pathog* 6, e1000905.

Sheppard SK, Colles F, Richardson J, *et al.* (2010a) Host Association of Campylobacter Genotypes Transcends Geographic Variation. *Applied and Environmental Microbiology* 76, 5269-5277.

Sheppard SK, Colles FM, McCarthy ND, *et al.* (2011a) Niche segregation and genetic structure of Campylobacter jejuni populations from wild and agricultural host species. *Molecular Ecology* 20, 3484-3490.

Sheppard SK, Dallas JF, MacRae M, *et al.* (2009a) Campylobacter genotypes from food animals, environmental sources and clinical disease in Scotland 2005/6. *International Journal of Food Microbiology* 134, 96-103.

Sheppard SK, Dallas JF, Strachan NJ, *et al.* (2009b) Campylobacter genotyping to determine the source of human infection. *Clinical Infectious Diseases* 48, 1072-1078.

Sheppard SK, dallas JF, Wilson DJ, *et al.* (2010b) Evolution of an agriculture-associated disease causing Campylobacter coli clade: evidence from national surveillance data in Scotland. In: *PLoS ONE*, p. e15708.

Sheppard SK, McCarthy ND, Falush D, Maiden MC (2008) Convergence of *Campylobacter* species: implications for bacterial evolution. *Science* 320, 237-239.

Sheppard SK, McCarthy ND, Jolley KA, Maiden MCJ (2011b) Introgression in the genus *Campylobacter*: generation and spread of mosaic alleles. *Microbiology*.

Sproston EL, Ogden ID, Macrae M, *et al.* (2011) Temporal variation and host association in the campylobacter population in a longitudinal ruminant farm study. *Appl Environ Microbiol* 77, 6579-6586.

Sproston EL, Ogden ID, MacRae M, *et al.* (2010) Multi-locus sequence types of Campylobacter carried by flies and slugs acquired from local ruminant faeces. *J Appl Microbiol* 109, 829-838.

Strachan NJ, Gormley FJ, Rotariu O, *et al.* (2009) Attribution of Campylobacter infections in northeast Scotland to specific sources by use of multilocus sequence typing. *Journal of Infectious Diseases* 199, 1205-1208.

Studahl A, Andersson Y (2000) Risk factors for indigenous campylobacter infection: a Swedish case-control study. *Epidemiology and Infection* 125, 269-275.

Thakur S, Gebreyes WA (2010) Phenotypic and genotypic heterogeneity of Campylobacter coli within individual pigs at farm and slaughter in the US. *Zoonoses Public Health* 57 Suppl 1, 100-106.

Waldenstrom J, Axelsson-Olsson D, Olsen B, *et al.* (2010) Campylobacter jejuni colonization in wild birds: results from an infection experiment. *PLoS ONE* 5, e9082.

Waldenstrom J, On SL, Ottvall R, Hasselquist D, Olsen B (2007) Species diversity of campylobacteria in a wild bird community in Sweden. *J Appl Microbiol* 102, 424-432.

Williams NJ, Jones TR, Leatherbarrow HJ, *et al.* (2010) Isolation of a novel Campylobacter jejuni clone associated with the bank vole, Myodes glareolus. *Appl Environ Microbiol* 76, 7318-7321.

Wilson DJ, Gabriel E, Leatherbarrow AJ, *et al.* (2009) Rapid evolution and the importance of recombination to the gastroenteric pathogen Campylobacter jejuni. *Mol Biol Evol* 26, 385-397.

Wilson DJ, Gabriel E, Leatherbarrow AJH, *et al.* (2008) Tracing the source of campylobacteriosis. *PLoS Genetics* 26, e1000203.

Zvelebil M, Dolukhanov P (1991) The transition to farming in eastern and northern europe. *Journal of World Prehistory* 5, 233-278.

4

The Generation of a Biodiversity Hotspot: Biogeography and Phylogeography of the Western Indian Ocean Islands

Ingi Agnarsson[1,2] and Matjaž Kuntner[2,3]

[1]*University of Puerto Rico, Puerto Rico,*
[2]*Department of Entomology, National Museum of Natural History,*
Smithsonian Institution,
[3]*Institute of Biology, Scientific Research Centre, Slovenian Academy of Sciences and Arts,*
[1,2]*USA*
[3]*Slovenia*

1. Introduction

The importance of islands in revealing evolutionary processes was highlighted already at the birth of evolutionary biology as a science (Darwin 1859; Darwin and Wallace 1858). Since the thrilling discoveries revealed by Darwin's work on the Galapagos (Darwin 1909) and Wallace's work in the Malay (Indonesian) archipelago (Wallace 1876), island biogeography has experienced an explosive growth. The discipline has provided many elegant examples of the evolutionary mechanisms involved in generating biodiversity, especially the interplay of geological processes and colonization and isolation (Emerson 2008; Gillespie, Claridge, and Goodacre 2008; Parent, Caccone, and Petren 2008; Ricklefs and Bermingham 2008). Islands have provided particularly strong insights into adaptive radiations (Camacho-Garcia and Gosliner 2008; Blackledge and Gillespie 2004; Cowie and Holland 2008; Gillespie and Roderick 2002; Losos and DeQueiroz 1997; Schluter 2000), the processes of colonization and extinction (Ricklefs and Bermingham 2008; Goldberg, Lancaster, and Ree 2011), the formation of species (Emerson 2008; Pickford et al. 2008; Gillespie and Roderick 2002; Schluter and Nagel 1995; Vences et al. 2009), and convergent evolution and formation of ecomorphs (Bossuyt and Milinkovitch 2000; Gillespie 2004, 2005; Losos 1988; Wildman et al. 2007; Burridge 2000; Rothe et al. 2011). Naturally, islands have also played a key role in revealing the causes and consequences of long distance dispersal, in particular, the ecological and evolutionary consequences of varying dispersal propensities of different lineages, and the evolutionary changes in dispersal propensity, such as the loss of dispersal ability following island colonization (Cowie and Holland 2006, 2008; Hedges and Heinicke 2007; Holland and Cowie 2006; Byrne et al. 2011; Bell et al. 2005; Darwin 1909; Clark 1994; Gillespie et al. 2012).

The field of phylogeography, in contrast, is much younger. Phylogeography—the study of phylogenetic data in a geographical context aimed to understand species distribution and diversity—is a rapidly expanding discipline that grew from a seminal paper by Avise and colleagues in 1987 (Avise et al. 1987). With the advent of easy sequencing, Avise foresaw a conceptual revolution in evolutionary biology: the boundary between population genetics (genetic studies within species) and phylogenetics (studies of the history of divergences among species) was dissolving (Avise et al. 1987; Avise 2000; Knowles and Maddison 2002; Hickerson et al. 2010), as is now the distinction between ecological and historical biogeography (e.g. Avise 2004; Ricklefs and Jenkins 2011). The genealogical history of a single locus, the "gene tree", is a common currency for both those biologists looking within species to study evolutionary dynamics of populations, and those looking across species to reconstruct and interpret phylogeny (Hickerson et al. 2010; Maddison 1997; Maddison 1995). Phylogeography was originally portrayed as the mitochondrial DNA link between these disciplines (Avise et al. 1987). However, phylogeography has since broadened immensely (for a recent review, see Hickerson et al. (2010)), with a current focus on gathering multiple gene trees. In animals, these usually represent mitochondrial and nuclear markers (Hare and Avise 1998; Charruau et al. 2011; Peters et al. 2007; Bryson, Garcia-Vazquez, and Riddle 2011; Fijarczyk et al. 2011; Rocha, Harris, and Posada 2011; Smykal et al. 2011; Witt, Zemlak, and Taylor 2011; Zhan and Fu 2011), and in plants, plastid and nuclear data (Buerki et al. 2011; Cuenca, Asmussen-Lange, and Borchsenius 2008; Manns and Anderberg 2011; Pokorny, Olivan, and Shaw 2011; Thiv et al. 2006). Modern phylogeography also incorporates other tools and disciplines, such as spatial analyses using GIS, and ecological niche modelling (Chan, Brown, and Yoder 2011; Raxworthy et al. 2007) and the emerging statistical coalescence theory (Maddison 1997; Knowles and Maddison 2002). Current development in next generation sequencing technology promises to complete the union of population genetics and phylogenetics: 'phylogenomics' will offer a wealth of data to unravel with increased confidence the histories of individuals, populations, and lineages at any level.

Phylogeography and biogeography are both sciences lying at this disappearing interface between genetic studies within and among species (Avise 2004; Avise et al. 1987). Both study genetic structure of populations in a geographical context and to both the concepts of climatological history, geographical barriers, and dispersal, are fundamentally important. Phylogeography could be defined simply as the phylogenetic analysis of data in a geographical context. However, the actual reach of phylogeography is much broader than such a definition implies. It aims to reveal the interplay of geological and climatological phenomena, life history traits, dispersal, and species distributions, in the generation of biodiversity. The data are typically genetic, for one thing because many data are available, and for another because they allow estimates of divergence dates. Conceptually, however, nothing rules out using morphological, behavioral, acoustic, chemical, and other heritable data that vary among individuals and populations to infer on phylogeography. Comparative phylogeography then aims to analyze data from multiple codistributed taxa and thus gets at broad questions about how historical global events (changes in geology, climate and so on) have shaped, and are shaping, the distribution and diversity of organisms on earth (Avise 2000; Hickerson et al. 2010). Codistributed taxa, especially those that are of comparable ages, are likely to have experienced similar geological and climatological histories and thus can offer insights into broader patterns than studies confined to single species/lineages.

Isolated archipelagos such as Hawaii and the Galapagos provide examples where cycles of evolutionary radiation have produced replicated patterns of endemic, often bizarre, forms

Cowie and Holland 2008; Dunbar-Co, Wieczorek, and Morden 2008; Parent, Caccone, and Petren 2008). Yet, the extreme isolation of these islands reduces the interplay between islands and continents — interchange is typically one-way (islands as sinks) and limited to rare chance dispersal events (Cowie and Holland 2008). Other archipelagos that are geologically closer to continental landmasses, such as the West Indies and the islands of the western Indian Ocean, can have very different geological and biological histories than highly isolated archipelagos (see, e.g. Leigh et al. 2007). The origin of such islands may be volcanic, or represent fragmentation from a continental landmass (such as Madagascar, and the Greater Antilles), and thus the origin and evolution of biota inhabiting these islands are expected to be more varied and dynamic (Ricklefs and Bermingham 2008). For example, colonization events are expected to be more frequent such that an island's diversity may not be dominated by adaptive radiations of a few lineages, but a mixture of many. Furthermore, especially when islands are close to continents, and/or very large, islands may function as a source of continent fauna, not just as sinks (Heaney 2007; Bellemain, Bermingham, and Ricklefs 2008; Bellemain and Ricklefs 2008). The biota of highly isolated oceanic islands is characterized by the near exclusive presence of excellent dispersers, some of which may have secondarily lost their dispersal ability, such as many spider lineages on Hawaii (e.g. Gillespie et al. 2012). In contrast, on archipelagos near continents, a complex mixture of good and poor dispersers is present, and the diversity and history of any given organismal lineage will strongly depend directly on its dispersal ability (Cowie and Holland 2006, Gillespie et al. 2012).

Our aim here is to review phylogeographic and biogeographic patterns of multiple, partially or fully codistributed taxa, to understand the broad strokes of the history of the terrestrial and freshwater biota of the land masses of the western Indian Ocean. First, we are interested in understanding the major shared biogeographical patterns in the region, and specifically how the dispersal abilities of different taxa affect those patterns. The basic prediction is that the better the dispersal abilities of a taxon, the less likely it is to share biogeographical patterns with many other taxa. Excellent dispersers, such as active day flying animals, can colonize islands stochastically practically from any direction, or source landmass (Gillespie et al. 2011). Poorer dispersers, on the other hand, are likely to show shared patterns that reflect geological history, climatological events, trade winds, direction of oceanic currents and other abiotic factors shaping species distributions. Second, we are interested in the impact of dispersal ability on patterns of species diversity to test a model presented in Agnarsson and Kuntner (in prep.). We predict that on near-continent archipelagos, poor dispersers will show patterns akin to organisms on isolated oceanic islands, that is: within island radiations on those islands where they occur as the result of speciation processes (e.g. Pearson and Raxworthy 2009, Gillespie et al. 2012). Radiations of poor dispersers will mostly occur on fragment islands, or larger islands (Leigh et al. 2007) relatively close to continents, but will not be present on most isolated volcanic islands. Excellent dispersers are expected to be found on most islands. However, radiations are not expected within islands, and may not occur among islands either if dispersal ability is sufficient to maintain among island gene flow (Kisel and Barraclough 2010, Gillespie et al. 2012). Thus, on near-continent archipelagos we predict the highest diversity of intermediate dispersers (Kuntner and Agnarsson 2011a-b; Agnarsson and Kuntner in prep.) — a diversity pattern that may be very general (Vilgalys and Sun 1994) — as those lineages have the opportunity to colonize even isolated volcanic islands, but may do so rarely enough such that gene flow is disrupted and among island genetic divergence begins to form immediately following colonization (Garb and Gillespie 2009, Gillespie et al. 2011). Intermediate dispersers therefore should be found on many islands and show divergence among them. They may also show divergences across barriers within the largest islands. In this chapter, we review various biogeographical studies on the terrestrial and freshwater biota of Indian Ocean islands to summarize biogeographical patterns in the western Indian Ocean and we test if diversification patterns across lineages follow the 'intermediate dispersal model' (Agnarsson and Kuntner In Prep).

The major islands of the western Indian Ocean, such as Madagascar, Comoros (including Mayotte), Réunion, Mauritius, Rodrigues, the Seychelles (including Aldabra), and Socotra (Figure 1), combine species richness with high endemicity resulting in high biodiversity. The region is one of Conservation International Biodiversity Hotspots (see http://www biodiversityhotspots. org/xp/hotspots/madagascar/Pages/default.aspx). Yet, the diversity of many lineages, in particular invertebrate animals, for example spiders, has barely begun to be studied in the region, and in general, the origin of the Indian Ocean biota, and mechanisms that have generated this diversity, are poorly understood and much debated (reviewed in Yoder and Nowak (2006), and Vences et al. (2009), see also Gage et al. (2011), Warren et al. (2010), and Kuntner and Agnarsson (2011a-b)). The origin and diversification of Indian Ocean lineages are challenging research topics because of the interplay of the complex geographic and geologic history of the islands, the varying dispersal abilities of local biota, and the proximity to the African continent. For example, Gondwanan vicariance dating back over 100 million years explains the origin of a portion of the diversity known on Madagascar (Briggs 2003), but not that found on the smaller, younger islands. The origin of those lineages that naturally occur there needs to be explained by Cenozoic dispersal from Africa, Madagascar, Asia, or Australasia (Yoder and Nowak 2006; Kohler and Glaubrecht 2010). The mixture of origins and the proximity of the islands combined provide a dynamic interaction and interchange of biota between continents and islands, which is quite different to the more isolated archipelagos, such as Hawaii (Gillespie, Claridge, and Goodacre 2008). Such mechanism likely represents an important force in the generation of biodiversity of the archipelago. Furthermore, a paradigm shift is underway in island biogeography, where molecular phylo/biogeographic studies are revealing an increasing importance of 'reverse colonization,' from islands to continents (e.g. Heaney 2007), in addition to the more commonly invoked model of colonization from continents onto islands. Some primary examples supporting this new paradigm come from the Indian Ocean, such as the diversification of chameleons on Madagascar followed by 'reverse colonization' of Africa, likely via rafting over the Mozambique Channel (Raxworthy, Forstner, and Nussbaum 2002). The islands of the Indian Ocean are thus not just a "museum of biodiversity" but also a cradle that has provided an important evolutionary source of diversity for the African continent. Indian Ocean taxa have provided a means to understand general evolutionary principles, including multiple text book examples of adaptive radiation, ranging from dung beetles (Orsini, Koivulehto, and Hanski 2007), to day geckos (Harmon et al. 2008), to chameleons (Raxworthy, Forstner, and Nussbaum 2002), and to lemurs (e.g. Martin 2000), to name but a few (see also Yoder and Nowak 2006; Kuntner and Agnarsson 2011a-b).

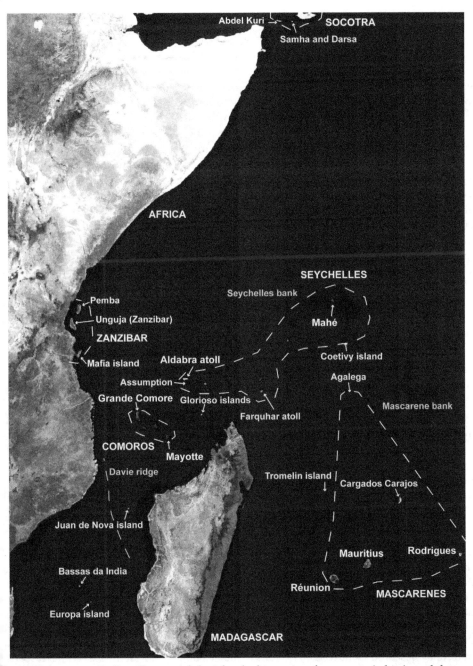

Fig. 1. The western Indian Ocean and the islands this review focuses on (white), and those briefly mentioned (yellow). Oceanic ridges and banks are indicated in green and their outlines can be seen in gray. Dotted white lines surround island archipelagos.

2. Paleogeology of the western Indian ocean

The geological history of the western Indian Ocean islands is complex, and not completely resolved. The western Indian Ocean contains both 'Wallacean' or 'fragment' islands such as Madagascar, Socotra, and the granite Seychelles that once formed a part of Gondwana, as well as 'Darwinian' or 'de novo' volcanic islands of varying ages, such as the Comoros, the coral Seychelles, and the Mascarenes. Sprinkled throughout the area are also many small to tiny coral islands and atolls that also are Darwinian, and some large continental Wallacean islands that have only recently been separated from Africa, such as Zanzibar. We here summarize the broad strokes in the geological history of the area. We focus on the patterns displayed by the larger offshore islands, while offering some remarks on other islands in the region. Very few comparative data are available on the biota of the many tiny islands scattered through the western Indian Ocean, such as Europa, Bassas da India, Juan de Nova, Glorioso, Agalega, Cargados Carajos, and Tromelin islands, labelled yellow in Fig. 1. Several studies have looked at bird and turtle populations on some of these, but little is otherwise known about their biota and their origins. However, they represent fascinating opportunities for biogeographical studies. The relatively large islands lying very close to Africa in the Zanzibar archipelago (25-60 km away from the mainland) were connected with Africa during the last ice age and are thus very recently separated from the mainland, except Pemba, which has been isolated for probably several million years. Socotra is an old and isolated Gondwanan fragment of the African continent, but very distant from the remaining islands and thus, at least in part, has a different biogeographical history.

2.1 Wallacean islands

'Wallacean islands', or 'fragment islands' are islands that once formed a part of a larger landmass, but that through tectonic plate activities and fragmentation of continental crust became secondarily isolated. In principle, such islands were 'born' with continental biota and ought to be species rich upon birth, with ecological space initially filled. Species richness then changes after fragmentation as a function of extinction, colonization, and in situ speciation (Gillespie and Roderick 2002). However, many geological Wallacean islands including the granite Seychelles are biologically better characterized as Darwinian as they have been submerged subsequent to their isolation, and thus most of the current biota will have arrived there via dispersal overwater after the emergence of the islands.

2.1.1 Madagascar

Madagascar is a large Wallacean island (~587,000 km²) separated from Africa by the Mozambique Channel, some 420 km at the narrowest point. Yoder and Nowak, and Masters and colleagues (Yoder and Nowak 2006; Masters, de Wit, and Asher 2006) offer detailed, but rather different, reviews of the complex geological history of Madagascar and its interaction with the rest of Gondwana (see also Leigh et al. 2007). Yoder and Nowak highlight recent phylogenetic evidence of numerous taxa and how results generally are at odds with predictions based on Gondwanan vicariance, rather pointing towards transoceanic dispersal during the Cenozoic as the primary biogeographical force. Masters and colleagues, however, highlight Gondwanan elements of the Malagasy fauna, and the possibility of land bridges allowing colonization after continental breakup over land rather than water. We will here build on these reviews by succinctly summarizing them, and point readers to Yoder and Nowak's and Masters et al. papers for specific details and references. However, we also highlight more recent findings, largely in support of the view of Yoder and Nowak, that offer new evidence for transoceanic dispersal mechanisms in the Cenozoic (Ali and Huber 2010).

Madagascar formed a part of eastern Gondwana, which also contained India, Antarctica, and Australia, while western Gondwana included Africa and South America (Briggs 2003). During 165 to 155 Ma the parts that would form eastern Gondwana drifted southward along the eastern African continent, and by 140 Ma a body of ocean already separated Madagascar from Africa at which point biotic exchange would have been dramatically reduced (see Yoder and Nowak (2006) and references therein). By 118 Ma, and possibly as early as 130 Ma, Madagascar reached its current position relative to Africa, already separated by the deep and wide Mozambique Channel. However, Madagascar remained connected to India for much longer. The timing of separation is debated, but sometime between 100 Ma and 87 Ma India plus the Seychelles block was separating from Madagascar, moving northeast. Thus the separation occurred gradually over millions of years, with the final separation of Madagascar and India frequently being cited as 88 Ma (Storey et al. 1995), but see (Rage 2003). The history of India and the connection between IndoMadagascar and the remainder of eastern Gondwana is complex, and seems far from settled. Some argue that India was already isolated by 130-125 Ma, while others suggest, based on recent fossil findings, that connections existed between IndoMadagascar and Antarctica and from the latter to S. America until as late as 80 Ma (see Yoder and Nowak 2006 and references therein). Such land bridges might explain the presence of elephant birds in Madagascar, as well as abelisaurid dinosaurs, and mammals with close affinities to S. America. Furthermore, there are several groups, especially non-volant mammals, whose history does not fit well the Gondwanan breakup as currently understood, but whose presence in Madagascar also cannot be readily explained by long distance dispersal (Masters, de Wit, and Asher 2006). Instead, Masters and colleagues suggest that several plateaus, banks, and fracture zones may have emerged periodically from the ocean and facilitated biotic interchange between Africa, India, and Madagascar (Masters, de Wit, and Asher 2006). For example, the distribution of certain taxa such as lemuriform primates, boine snakes, hyperoliid tree frogs, iguanid lizards, and certain plant groups, has been suggested to imply a more recent connection between India and Madagascar. This prompted the "Lemurian Stepping-stones" hypothesis (Schatz 1996; Steenis 1962). The hypothesis proposes that dispersal to Madagascar from India and further from Asia was facilitated by existing islands in the Eocene (56-34 Ma), close to the time India assumed its current position. At that time, a global drop in sea levels may have exposed the Seychelles bank and the Mascarene and Chagos/Laccadive plateaus and these, in turn, may have served as 'stepping stones' facilitating the exchange of biota between India and Madagascar (Rage 2003). Similarly, the Davie fracture zone lying between Africa and Madagascar is hypothesized by some to have partially emerged from the ocean during episodes some 45–26 million years ago (Masters, de Wit, and Asher 2006). However, there are fundamental problems with such land bridge arguments. First, most of these ridges would have, at best, partially emerged, creating island chains which still would require, albeit shorter distance, transoceanic dispersal. Second, the existence of a solid land-bridge, for example between Africa and Madagascar makes specific predictions, among which are that many different groups of taxa should have arrived, while only a few did, and the taxa that did arrive should have done so more or less simultaneously, which is also not the case (Table 1, Ali and Huber (2010); Yoder and Nowak (2006)).

Kingdom	Taxon	Pattern summary	Estimated dispersal events	Estimated regional diversity	Reference
animals	*Hemicordulia* dragonflies	Likely two colonization events of W Indian Ocean islands and Africa from Asia assisted by westward storms, but precise routes unknown in the absence of phylogeny	2	5	Dijkstra (2007)
animals	Scops owls (*Otus*)	Multiple colonizations from Indo-Malaya (order undetected but estimated at 3.6 ma), and one back colonization of Asia; Indian Ocean radiation at 2.5 ma; Pemba colonization separate from Africa at 1.7	3	7	Fuchs et al. (2008)
animals	Parrots	Likely colonization from India via stepping stones	1	9	Hume (2007)
animals	Parrots	Dispersals twice from Australasia to Madagascar, once from Madagascar to Africa and once from Australasia to Africa	4	7	Schweizer et al. (2010)
animals	Hawksmoth genus *Hyles*	Dispersal from South America to Australia and Madagascar	1	2	Hundsdoerfer et al. (2009)
animals	Oscine passerine birds	Dispersal from Australia to Africa via Indian Ocean islands, then from Africa to other continents	1	n/a	Jonsson and Fjeldsa (2006)
animals	Golden orb spider (*Nephila inaurata*)	Africa to Madasacar to Mascarenes	10+	1	Kuntner and Agnarsson (2011)
animals	*Colotis* butterflies	Multiple dispersals from Africa to Madagascar	6	7	Nazari et al. (2011)
animals	*Pteropus* fruit bats	Independent colonizations: Asia to Comoros/Pemba, Asia to Rodrigues and Asia to all other W Indian Ocean islands	8	8	O'Brien et al. (2009)

Kingdom	Taxon	Pattern summary	Estimated dispersal events	Estimated regional diversity	Reference
animals	Drongos (Dicruridae)	Asia to Africa, from there to Indian Ocean islands and back to Asia; Madagascar to Mayotte, Madagascar to Aldabra, Madagascar to Comoros	6	4	Pasquet et al. (2007)
animals	Carpenter bees (Ceratinini)	Two or three dispersals from Africa to Madagascar between 25 and 9 ma	2	?	Rehan et al. (2010)
animals	Allodapine bees	Dispersal events from Africa via Indian Ocean to Australia, and from Africa to Madagascar	2	35	Schwarz et al. (2006)
animals	magpie robins (Copsychus)	Asia to Madagascar	1	2	Sheldon et al. (2009)
animals	Canthonini and Dichotomiini (Scarabaeinae)	Dispersal events from Africa to Eurasia and on to America and Africa to India, SE Asia and to Australia	n/a	n/a	Sole and Scholtz (2010)
animals	Scarabaeini beetles	Africa to Madagascar	1	3	Sole et al. (2011)
animals	White-eye songbirds (Zosterops)	Complex history with an early dispersal to W Indian Ocean and to Africa from Asia and a later one from Africa to Comoros, then to other W Indian Ocean	8	15	Warren et al. (2006)
animals	Parrots	Australasian origin, speciation though vicariance, then dispersal to e.g. Madagascar from Australia	1	n/a	Wright et al. (2008)
animals	Starlings and mynas (Sturnus, Acridotheres)	Asia to Reunion (for Fregilupus varius), a split about 4 ma	1	6	Zuccon et al. (2008)
animals	Rattus rattus	Introduced, likely each island separately	n/a	1	Tollenaere et al. (2010)
animals	Bactrocera cucurbitae (Diptera: Tephritidae)	Introduced	0	1	Virgilio et al. (2010)
animals	Anelosimus spiders	Two independent colonizations of Madagascar/Comoros from Americas and Americas or Africa, both less than 10 ma	4	12	Agnarsson et al. (2010)

Kingdom	Taxon	Pattern summary	Estimated dispersal events	Estimated regional diversity	Reference
animals	*Mormopterus* bats	Inconclusive	?	2	Goodman et al. (2008)
animals	*Rousettus* bats	Inconclusive; rare dispersal over 300 km leading to diversification	2	2	Goodman et al. (2010)
animals	*Chaerephon* bats	Bats maintain gene flow within Comoros, but not further	1	2	Goodman et al. (2010)
animals	Hermit spiders (*Nephilengys*)	Inconclusive, either Africa to W Indian Ocean islands or vice versa	0	1	Kuntner (2007)
animals	Hermit spiders (*Nephilengys*)	Africa to Madagsacar, then to Mascarenes and Madagascar to Mayotte	3	4	Kuntner and Agnarsson (2011)
animals	*Drosophila mauritiana*	Gene flow from Mauritius to Rodrigues, but not vice versa	1	1	Legrand et al. (2011)
animals	*Triaenops* bats	At least two dispersals from Africa to Madagascar	2	4	Russell et al. (2008)
animals	*Scotophilus* bats	Two independent colonizations of Madagascar from Africa	2	4	Trujillo et al. (2009)
animals	*Miniopterus* bats	Madagascar to Comoros at least twice with limited or no subsequent gene flow	below	below	Weyeneth et al. (2008)
animals	*Miniopterus* bats	Madagascar to Comoros 180000 ya	2	2	Weyeneth et al. (2011)
animals	*Zaprionus* (Diptera)	Asia to Africa via Indian Ocean islands	2	7	Yassin et al. (2008)
animals	Geckos, skinks	Independent colonizations from Mauritius	2	3	Arnold and Bour (2008)
animals	*Leiolopisma* skinks	Single dispersal event to Mauritius from Australia, then to Reunion	2	3	Austin and Arnold (2006)
animals	Slit-eared skinks	Madagascar (or Africa) to Mauritius, to Reunion	2	3	Austin et al. (2009)
animals	Haemadipsidae leeches	Not vicariant	3	8	Borda and Siddall (2010)
animals	Ranid frogs	Vicariant - apparently, but dispersal to Mayotte	1	130	Bossuyt et al. (2006)
animals	Freshwater crabs	Multiple colonizations from Africa and Eurasia (Socotra)	below	below	Cumberlidge (2008)

Kingdom	Taxon	Pattern summary	Estimated dispersal events	Estimated regional diversity	Reference
animals	Seychellum alluaudi freshwater crab	Africa to Seychelles	below	below	Daniels (2011)
animals	Freshwater crabs	Dispersal to Madagascar and Seychelles	3	15	Daniels et al. (2006)
animals	Homopholis and Blaesodactylus gekkos	Africa to Madagascar	2	4	Greenbaum et al. (2007)
animals	Day geckos (Phelsuma)	Independent colonizations from Madagascar to Seychelles, Comoros, Mascarenes	10	49	Harmon et al. (2008)
animals	Day geckos (Phelsuma)	Colonizations of Mascarenes and Pemba from Madagascar	above	above	Raxworthy et al. (2007)
animals	Pristionchus pacificus nematode	Multiple independent invasions	?	1	Herrmann et al. (2010)
animals	Gecarcinucoidea freshwater crabs	African origin, stepping stone dispersal via W Indian Ocean islands towards Asia	5	19	Klaus et al. (2006)
animals	River snails Pachychilidae	Cenozoic dispersal	1	5	Kohler and Glaubrecht (2010)
animals	Oplurin iguanas	No diferentiation between Comoro and Madagascar	1	7	Munchenberg et al. (2008)
animals	Stenophis and Lycodryas snakes		8	82	Nagy et al. (2010)
animals	podocnemid turtles	Vicariant	0	1	Noonan and Chippindale (2006)
animals	iguanid lizards	Vicariant	0	7	Noonan and Chippindale (2006)
animals	Boid snakes	Vicariant	0	3	Noonan and Chippindale (2006)
animals	Agamid and chamaeleonidae	Vicariant Madagascar-Indian	0		Okajima and Kumazawa (2010)
animals	Plated lizards (Gerrhosauridae)	Likely from Africa to Madagascar, followed by radiation	1	19	Raselimanana et al. (2009)
animals	Cetartiodactyla	Single dispersal from Africa	1	3	Yoder and Nowak (2006)

Kingdom	Taxon	Pattern summary	Estimated dispersal events	Estimated regional diversity	Reference
animals	Eulipotyphla	Single dispersal from Africa	1	1	Yoder and Nowak (2006)
animals	Rodents	Single dispersal from Africa	1	27	Yoder and Nowak (2006)
animals	Carnivora	Single dispersal from Africa	1	9	Yoder and Nowak (2006)
animals	Afrotheria	Single dispersal from Africa	1	30	Yoder and Nowak (2006)
animals	Perissodactyla	Present in Africa, no dispersal to the islands	0	0	IUCN
animals	Pholidota	Present in Africa, no dispersal to the islands	0	0	IUCN
animals	Onychoprion	Dispersal among islands frequent enough to maintain gene flow	frequent	1	IUCN
animals	Pantala	Dispersal among islands frequent enough to maintain gene flow	frequent	1	IUCN
animals	Primates	Suggests vicariant origin of primates	below	below	Heads (2010)
animals	Primates	Single dispersal from Africa	1	94	Yoder and Nowak (2006)
animals	Coastal lizards (Cryptoblepharus)	Colonization by transoceanic dispersal (from Australia or Indonesia) to Madagascar, from there to Africa, Comoros islands (twice), and Mauritius	5	Unresolved	Rocha et al. (2006)
animals	Mabuya lizards (Scincidae)	West to East across Comoros	4	14	Rocha et al. (2010)
animals	Streptaxid land snails	Africa to Madagascar, Africa to Seychelles, Africa to Mascarenes	12	240	Rowson et al. (2011)
animals	Crayfishes (Parastacidae)	Vicariant (Gondwana)	0	6	Toon et al. (2010)
animals	Archaius (Calumma) chameleons	Africa to Seychelles	1	1	Townsend et al. (2011)
animals	Polystomatid flatworms (amphibian parasites)	Two independent colonizations of Madagascar via amphibian hosts, maybe one from Asia the other from Africa	2	13	Verneau et al. (2009)
animals	Blindsnakes (Typhlopidae)	Vicariant (Gondwana)	1	14	Vidal et al. (2010)
animals	Giant pill-millipedes (Sphaerotheriida)	Vicariant (Madagascar)	0		Wesene and VandenSpiegel (2009)

Kingdom	Taxon	Pattern summary	Estimated dispersal events	Estimated regional diversity	Reference
animals	Giant pill-millipedes (Arthrosphaeridae)	Vicariant (Gondwana)	0	60	Wesener et al. (2010)
animals	Caecilian amphibians	Vicariant (Seychelles)	0	6	Zhang and Wake (2009)
animals	Hemidactylus geckos	Overocean dispersal to Socotra from northern Africa, and from Arabia	2	11	Carranza and Arnold (2006)
animals	Chamaeleo monachus	Vicariant	0	1	Macey et al. (2008)
plants	Coffea (Rubiaceae)	Africa to Grand Comore	below	below	Anthony et al. (2010)
plants	Coffea (Rubiaceae)	See Wikstrom et al. (2010) Africa to Madagascar, Asia to	below	below	Maurin et al. (2007)
plants	Coffee family (Rubiaceae)	Madagascar, Madagascar to Comoros, Madagascar to Seychelles, Madagascar to Mascarenes	30	803	Wikstrom et al. (2010)
plants	Chrysophylloideae (Sapotaceae)	Africa to Madagascar THREE times	4	40	Bartish et al. (2011)
plants	Lilaeopsis (Apiaceae subfamily Apioideae)	Likely a recent introduction	0	1	Bone et al. (2011)
plants	Vanilla	Overocean dispersal to Reunion	3	10	Bouetard et al. (2010)
plants	Sapindaceae (Molinaea, Neotina, Tina, Tinopsis)		4	100	Buerki et al. (2011)
plants	Palm tribe Chamaedoreeae	Long distance dispersal (absent from Madagascar and Africa)	3	5	Cuenca et al. (2008)
plants	Scaly tree ferns	3 independent colonizations followed by speciation	8	56	Janssen et al. (2008)
plants	Hibisceae (Malvaceae)	Multiple colonizations of Madagascar	5	86	Koopman and Baum (2008)
plants	Fern genus Platycerium	Vicariant	4	6	Kreier and Schneider (2006)
plants	Euphorbs	Colonization of Madagascar/Africa from Asia	1	11	Kulju et al. (2007)
plants	Dombeyoideae (Malvaceae)	At least 5 colonizations from Madagascar	15	273	Le Pechon et al. (2009)
plants	Dombeyoideae (Malvaceae)	Multiple Madagascar to Mascarenes	above	above	Le Pechon et al. (2010)

Kingdom	Taxon	Pattern summary	Estimated dispersal events	Estimated regional diversity	Reference
plants	tropical Anagallis (Myrsinaceae)	Africa to Madagascar twice	2	10	Manns and Anderberg (2011)
plants	Hernandiaceae	Dispersal to Madagascar from Asia, and to Reunion from Pacific	2	4	Michalak et al. (2010)
plants	Angraecoid orchids	Multiple from Madagascar	20	321	Micheneau et al. (2008)
plants	Calyptrochaeta moss	Africa to islands?	3	1	Pokorny (2011)
plants	Monimiaceae	Australasia to Indian Ocean islands	6	63	Renner et al. (2010)
plants	Lomariopsis ferns	Madagascar clade includes nested African species	3	7	Rouhan et al. (2007)
plants	Cucurbitaceae	13 independent colonizations fo Madagascar from Africa	14	90	Schaefer et al. (2009)
plants	Ferns (Lindsaeaceae)	Unclear	n/a	31	Lehtonen et al. (2010)
plants	Warneckea (Melastomataceae)	Madagascar from Africa, possibly Mauritius from Madagascar, the authors are not interested in biogeography	1	20	Stone and Andreasen (2010)
plants	Aerva (Amaranthaceae)	Eritreo-Arabian colonization of Socotra	5	8	Thiv et al. (2006)
plants	Baobab Adamsonia digitata	Natural and human dispersal ot Madagascar from Africa	1	1	Tsy et al. (2009)
plants	Diospyros (Ebenaceae)	Colonization of Mauritius, followed by Reunion and Rodrigues	2	14	Venkatasamy et al. (2006)
plants	Euphorbia	Africa to Madagascar several times	5	110	Zimmermann et al. (2010)
plants	Echidnopsis (Apocynaceae)	Dispersal from E. Africa to Socotra	1	5	Thiv and Meve (2007)
plants	Thamnosma (Rutaceae)	Africa to Socotra	1	2	Thiv et al. (2011)

Table 1. A summary of recent literature on the phylogeography and biogeography of the western Indan Ocean islands. Estimated dispersal events represent our own interpretation of the data presented in the reference paper. Estimated regional diversity represents our own attempts to estimate the number of species of each clade in the region, this number in many cases does not come from the referenced paper, but from various external sources. Estimates in many cases are approximate due to limited knowledge and scattered information. Exemplar studies of Indian Ocean taxa that provide phylogeographic context. See "2. Review of studies and patterns" for details on filtering the literature search results.

Any dispersal model needs to consider not only the geological separation of landmasses, but also the direction of dispersal agents, such as trade winds and oceanic currents, through history. After the separation of Madagascar with Africa and India, oceanic currents are thought to have been favourable to dispersal on rafts from Africa to Madagascar, however, in the present day, currents go in the opposite direction, and are more favourable for colonization from Madagascar to Africa. How long the present day currents have persisted is debated, and some have suggested they may have prevailed for as much as 50 Ma (Masters, de Wit, and Asher 2006 and references therein), thus drastically reducing opportunities for colonization of Madagascar from Africa. However, an exciting new paleo-oceanographic modelling study concluded that strong oceanic currents did flow from eastern Africa to Madagascar during the Palaeogene, with occasional currents strong enough to cross the Mozambique channel in less than a month (Ali and Huber 2010). Starting soon after the great faunal turnover at 65 Ma (see below) and lasting until about 20 Ma (Ali and Huber 2010), these favourable currents coincide in time with the arrival of non-volant mammals as reconstructed by dated phylogenies: lemuriform primates (60-50 Ma), tenrecs (42-25 Ma), carnivorans (26-19 Ma), and rodents (24-20 Ma) (Poux et al. 2005; Yoder and Nowak 2006). Under this model, the timing of currents shifting towards their present day flow, from Madagascar to Africa, also fits observation as this time more or less marked the end of non-volant mammal arrival. Furthermore, this scenario does not predict the simultaneous arrival of groups of organisms, but requires only the very rare occurrence of successful overwater dispersal (once per several million years) under relatively favourable conditions.

Though the geological history of Madagascar and related landmasses and oceanic plateaus is complex, Madagascar is clearly Wallacean in its geological origin. However, catastrophic geological events may have wiped out a portion of the Gondwanan fauna and flora on Madagascar about 65 Ma, through meteor impact in India and the subsequent Deccan volcanism (see below).

2.1.2 The granite Seychelles

The Seychelles is an archipelago of 155 islands, islets, and rocks, lying about 1.100 km northeast of Madagascar and 1.500 km east of Africa, with a total land area of only 220 km². About 42 of these are granite islands that are thought to be continental fragments that were once part of the Indo-Madagascan block that broke away from India during its passage northeast to Asia (Rabinovitz, Coffin, and Falvey 1983). These islands may have been isolated for upwards of 65 Ma (Plummer and Belle 1995), however they have all been mostly or completely submerged at various times with changes in sea level presumably wiping out most of the terrestrial biota. Thus from a biogeographical standpoint the granite Seychelles have many characteristics of Darwinian islands (Rabinovitz, Coffin, and Falvey 1983) with most of the biota having arrived via dispersal. Even some 'classic' examples of vicariant taxa, such as freshwater crabs, are now thought to have arrived via transoceanic dispersal (Cumberlidge 2008; Daniels 2011; Daniels et al. 2006). The notable exceptions are burrowing animals: caecilians (Zhang and Wake 2009) and sooglossid frogs (Biju and Bossuyt 2003), and possibly some ferns (Lehtonen et al. 2010), that appear to have persisted on the islands even during the periods of high sea levels. Some taxa, such as the sooglossid frogs, show genetic divergence among the granite Seychelles e.g. (Van der Meijden et al. 2007), highlighting the need for thorough sampling and biogeographical analyses of these islands.

2.1.3 Socotra archipelago

The Socotra archipelago (Yemen) consists of four islands, Abdel Kuri, Samha, Darsa, and by far the largest island—Socotra—measuring 3,625 km². The archipelago is isolated (240 km from nearest point in Africa, 380 km from Arabia) having separated from the Arabian peninsula probably over 30 Ma, in the Oligocene (Thompson 2000). The Arabian peninsula itself started to separate from northern Africa about 60 Ma (Thompson 2000) eventually crashing into the Eurasian plate about 15 Ma. While the islands are separated by shallow seas and may have come in contact with each other during episodes of low sea levels, they are separated from the continents by a deep trench (Busais 2011). Thus the biota of Socotra is characterized by high endemism. Of the large flora of Socotra, counting over 800 species, some 37% are endemic (Miller and Miranda 2004). Socotra is home to over 30 lizard species, 80% of which are endemic (Rösler and Wranik 2004), and also hosts a range of other endemic animals including geckos (Busais 2011), chameleons, colubrid snakes, and freshwater crabs (Cumberlidge 2008). The smaller islands also have endemics, such as the Abdel-Kuri Sparrow, and the day gecko *Hemidactylus forbesii* (Carranza and Arnold 2006). Some of Socotra's biota is thought to be vicariant, including the Socotran chameleon (Macey et al. 2008), and possibly some geckos (Arnold 2009; Gamble et al. 2008), while other elements are thought to have arrived via transoceanic dispersal from Eurasia such as freshwater crabs (Cumberlidge 2008; Daniels et al. 2006; but see Shih, Yeo, and Ng (2009)) colubrid snakes, and various plants (Thiv et al. 2006). Yet others have arrived from Africa (Thiv and Meve 2007), and day geckos apparently arrived once from Arabia and once from Africa (Carranza and Arnold 2006).

2.1.4 Zanzibar archipelago

The Tanzanian Zanzibar archipelago consists of over 40 islands and islets, of which three are by the far largest: Unguja—collectively referred to as Zanzibar—(2,461 km²), Pemba (980 km²), and Mafia island (394 km²). All are parts of continental Africa, and Unguja and Mafia are separated from the continent only by 25-30 km stretch of shallow ocean. These islands have therefore periodically been connected to continental Africa during low sea levels, with the last connection as recent as 10.000 Ya (Kingdon 1989). Thus their biota is mostly shared with the continent, with notable exceptions as noted below. Pemba differs in being a little further away (about 55 km) and is separated from the continent by a very deep trench, such that it has been isolated at least for some million years (Kingdon 1989). Thus Pemba has several well-known endemics such as the Pemba flying fox (Robinson et al. 2010), a *Mops* bat (Stanley 2008), Pemba scops-owl (Fuchs et al. 2008), Pemba white-eye (Vaughan 1929), and four other bird species, or subspecies (Vaughan 1929; Catry et al. 2000), a day gecko (*Phelsuma parkeri*) (Rodder, Hawlitschek, and Glaw 2010), a damselfly (Dijkstra, Clausnitzer, and Martens 2007), and several others. Coastal lizards have unique haplotypes on Pemba, differing from those found on the other islands, and those shared between, Unguja and the Tanzanian coast, which is consistent with relatively long island isolation (Rocha et al. 2006). Freshwater snails may have an endemic species on the archipelago, and also show genetic divergence among Pemba and Mafia islands (Stothard, Loxton, and Rollinson 2002). Land molluscs have several endemic species on the archipelago but apparently no more on Pemba than Unguja (Rowson 2007; Rowson, Warren, and Ngereza 2010), though DNA evidence is still lacking. Unguja has other endemics, the most striking perhaps the now presumed extinct Zanzibar leopard (a subspecies of *Panthera pardus*) and the Zanzibar red colobus

Nowak and Lee 2011), but others include a species of frog (Msuya, Howell, and Channing 006), crustaceans (Kensley and Schotte 2000; Olesen 1999) and other taxa. Nevertheless, the proximity with Africa means that most of the Zanzibar archipelago biota, including Pemba, s shared with Africa (e.g. (Clausnitzer 2003; Kock and Stanley 2009)), even among relatively poor dispersers such as freshwater crabs, species are shared with Africa (Cumberlidge 2008).

.2 Darwinian islands

Darwinian islands provide a clean slate to colonists where species richness increases initially through immigration and later through formation of neo-endemics. For highly isolated slands, and for large islands, adaptive radiation may fill empty niches (Gillespie and Roderick 2002; Gillespie 2004). As noted above, some islands that are Wallacean in their geographical origin, have subsequently become mostly biologically Darwinian, for example due to periodic submergence with changing sea levels. The larger and older Darwinian slands are generally volcanic in origin, and island chains such as the Comoros and the Mascarenes have formed as tectonic plates move over a hot spot (Duncan and Storey 1994; Emerick and Duncan 1982), a common pattern of island chain formation.

.2.1 The Comoros archipelago

The Comoros are a group of volcanic islands a little over 300 km NW of Madagascar and with a total land area of about 2200 km². The oldest island, Mayotte, is in the east, closest to Madagascar, with successively younger islands to the west , Grand Comore being the youngest and closest to Africa. Grande Comore is thought to be the result of very recent volcanic activity, some 0.01–0.5 Ma. Mayotte consists of two volcanoes that emerged from the ocean, probably a little less than 10 Ma (Audru et al. 2006, 2010). Estimates of the exact dates of the islands vary. In the geological atlas of Africa (Schlüter 2006), the island ages are given as 0.01 Ma (Gran Comore), 3.9 Ma (Anjouan), 5.0 Ma (Moheli), and 7.7 Ma (Mayotte). Nougier et al. (1986) estimate that the Comoros started forming around 10-15 Ma, such that land area may have been available for colonization for over 10 Ma. The terrestrial biota arrived mostly via overwater dispersal from Madagascar and from mainland Africa. Due to the relatively old age of the islands and their isolation, they are home to numerous endemic species of animals and plants, including 16 species of birds, two chiropterans, several *Phelsuma* day geckos, two *Furcifer* chameleons, other reptiles and amphibians and many invertebrates (Harris and Rocha 2009).

.2.2 The Mascarene Islands

The Mascarene Islands, the principal of which are Réunion, Mauritius, and Rodrigues, lie about 700, 900, and 1500 km east of Madagascar, respectively, and have a total land area of 4500 km² including a few small offshore islands. All are of volcanic origin and have never been connected to any other landmasses (McDougall and Chamalaun, 1969), their biota thus necessarily has arrived via transoceanic dispersal. The age of the islands has long been debated, especially that of Rodrigues, which traditionally has been thought to be the youngest island at 1.5 Ma (McDougall and Chamalaun, 1969), but recent evidence favours the contrary conclusion, that it is the oldest island (Thébaud et al. 2009). Thus, most authors now seem to agree that Réunion is the youngest island of the Mascarenes. Age estimates are often given at some 2-3 Ma (Deniel, Kieffer, and Lecointre 1992), but it may be as old as 5 Ma

(Gillot, Lefèvre, and Nativel 1994). Mauritius is estimated to have emerged some 8-10 Ma (McDougall and Chamalaun, 1969), with the major components of the island being formed about 6 Ma. Rodrigues is then thought to be at least as old as Mauritius, or 8–15 My (Anon 1998; Cheke and Hume 2008, Thébaud et al. 2009). Geologically, the Mascarene Plateau is as old as 40 Ma (Morgan 1981), so the possibility exists that a portion of the archipelago's land area was above ocean surface even earlier than 10 Ma. Islands now submerged are thought to have played an important role in the biogeographical history of some organisms in the region, such as the dodo, day geckos, and Laurales and palms, where colonization of the archipelago may have occurred before the current islands arose from the sea (Austin, Arnold, and Jones 2004; Cuenca, Asmussen-Lange, and Borchsenius 2008; Shapiro et al 2002; Renner et al. 2010).

2.2.3 The coral Seychelles and Aldabra atoll

As outlined above, the Seychelles consist of 155 islands and islets, 42 of which are ancient Wallacean fragments of Gondwana—the granite islands—and the remainder are coral islands and atolls. These coral islands vary greatly in age and size, with Aldabra Atoll being one of the oldest with proto Aldabra dating to the late Pleistocene (i.e. less than 2.5 Ma), the largest (155 km² land area, 224 km² lagoon), most isolated (425 km NW of Madagascar, the nearest major landmass), and mostly undisturbed by humans. All the coral islands have in common that they have been submerged periodically through changes in sea level. For example, the most recent emergence of Aldabra Atoll from the ocean occurred about 125,000 years ago (Warren et al., 2005). Yet, Aldabra has some endemic biota, including the Aldabra drongo and the now presumed extinct Aldabra warbler (IUCN 2011), and several endemic plants. For the other smaller coral islands the frequency of submergence and the timing of the most recent emergence relate to island size and, especially, altitude. Regardless, all coral islands in the region can be considered as recent Darwinian islands from the perspectives of terrestrial biogeographical and phylogeographical studies.

2.2.4 The 'minor' islands of the western Indian Ocean

This review focuses on the larger islands of the Indian Ocean, mainly because their biota is in general much better known. There are, however, scattered throughout the western Indian Ocean a number of highly isolated small islands, most of which are home to breeding colonies of oceanic birds, and turtles. Notable among these are, in the Mozambique Channel, Europa island, Bassas da India, and Juan de Nova island (Fig. 1). North of Madagascar are Glorioso island, and some of the furthest outlying coral Seychelles, such as Assumption (close to Aldabra), and Farquhar atoll (Fig. 1). In between the granite Seychelles and the three main Mascarene islands are further highly isolated islands: Coetivy island, one of the outlying Seychelles, Agalega and Cargados Carajos, two outlying Mascarene islands, and then Tromelin island sitting between Madagascar, and the Mascarenes and Mascarene bank (Fig. 1). Though all of these are interesting from a biogeographical standpoint, and indeed some are known to harbour endemic forms, few molecular studies have included taxa from these islands such that their biogeographical history is in general poorly known; for information on many of these see Stoddart (1970), Stuart and Adams (1990), Caceres (2003). Europa island (28 km²), about 300 km from Madagascar and 500 km from the African coast is an important breeding ground for sea turtles and oceanic birds. An endemic 'race' of the white tailed tropicbird occupies the island, and Europa populations of some other birds species appear to have little or no gene flow with other populations (WWF, 2008). The island also is host to subspecies such as Europa white eye and a snake-eyed skink. A few invertebrate species may be endemic, including the little known termite *Neotermes europae*, cockroach *Elliptorhina lefeuvri*, and leafcutter bee *Megachile pauliani* (see http: // lntreasures. com / fsal. html). Interestingly, breeding populations of green turtles are rather unique in Europa compared to the rest of the Indian ocean, and differ, for example from populations on Juan de Nova island (Bourjea et al. 2007). **Bassas da India** is a small atoll (~ 80 km² including lagoon) about 100 km north of Europa, that is mostly underwater during high tide and thus supports no vegetation, nor terrestrial fauna. **Juan de Nova island** is a tiny island (4 km²), home to a large colony of scooty terns (Jaquemet, Le Corre, and Quartly 2007) and has a unique genetic stock of green turtle (Bourjea et al. 2007). Of notable terrestrial fauna is a possibly endemic species of skink, *Cryptoblepharus caudatus*, and ant species shared with Madagascar *Camponotus hova*. **Glorioso islands,** consisting of five islands totalling about 5 km² in size, is home to several species of oceanic birds and a possibly endemic cone shell (*Conus veillardi*). **Assumption** is a small island (11 km²) close to the Aldabra atoll that is home to the endemic subspecies Assumption Island day gecko (*Phelsuma abbotti sumptio*), and a number of widespread oceanic birds. **Farquhar atoll** consists of a group of small islands that, although encircling a large lagoon, have a very small total land area (7.5 km²). The atoll is one of the few breeding places of Roseate terns, but both flora and fauna consist mostly of widespread species lacking endemics, though both are poorly known (Stoddart 1970; Stuart and RJ 1990). **Coetivy island** is a small sand cay (9.3 km²), apparently lacking endemics (Stuart and RJ 1990). **Agalega** consists of two islands with a total land area of 24 km² and is home to an endemic subspecies of the Réunion day gecko, (*Phelsuma borbonica agalegae*) (Cheke 2010; Rocha et al. 2009), **Cargados Carajos** is a group of tiny islets with a total land area of 1.3 km² and no terrestrial endemism. **Tromelin** is a tiny rock (0.8 km²), but is a host to some apparently endemic insects, such as the homopteran *Pulvinaria tromelini* (Stoddart 1970), and is the stronghold breeding place of a western Indian Ocean endemic subspecies of terns (*Sula dactylatra melanops*).

3. Review of biogeographical and phylogeographical studies and patterns

We present a review of studies on various terrestrial and freshwater taxa occupying the larger western Indian Ocean islands. The review does not aim to be exhaustive, but is explicitly exemplary in its approach, focusing on recent literature. For our review of patterns we focus on the islands that have been best studied, while highlighting the many fascinating isolated tiny islands that deserve research attention (Fig. 1). We survey in particular detail the literature published since the review of Madagascar biogeography (Yoder and Nowak 2006). In Web of Science we used the following literature search criteria: (Biogeography AND Madagascar OR Comoros OR Reunion OR Mauritius OR Rodrigues OR Socotra OR Seychelles OR Zanzibar) AND Year Published=(2006-2011); the hits were then refined by Web of Science categories: (Zoology OR Ecology OR Plant sciences OR Evolutionary biology) and by document type (Article OR Review), then further filtered by hand to retain only relevant works. The resulting Table 1 lists all those works that treat non-marine Western Indian Ocean taxa explicitly and contain information on the phylogenetic relationships of those taxa. For these taxa, we extracted a biogeographically useful summary of the pattern of colonization—those we tally up across all taxa and summarize in Fig. 2—and diversification in the archipelago. We also collected information on genetic divergence among landmasses, and estimated timing of colonization events, however such data were available in far too few studies for any comparative analyses, thus these results are not included in Table 1. Additionally, we discuss selected studies taken from older literature, and from the reviews of Yoder and Nowak (2006) and Masters, de Wit, and Asher (2006). We note that the sampled histories have a strong bias towards Madagascar, as many more studies are published that include taxa from Madagascar than any other island in the western Indian Ocean. Hence, when tallying independent dispersal histories, a comparison of absolute numbers is not very informative. Instead we interpret the main patterns of colonization of each archipelago and summarize observed histories in Fig. 2. We also attempt to estimate the relative role of vicariance, by counting putative vicariant histories and divide them by the total number of observed histories (total number of colonization events or vicariance events across independent lineages). However, this is necessarily a crude estimate, both because we look only across recent studies, and because many of the putative vicariant histories yet lack testing through molecular dating. We start this review with summarizing examples from various relatively well studied groups of animals and plants, and then move to studies on spiders, mostly from our own work. We summarize general patterns observed for each taxon, and then attempt to draw broad and general conclusions about Indian Ocean biogeography across taxa, the common colonization routes, and the general role of dispersal abilities in shaping both distributional and diversification patterns.

3.1 Biogeographical patterns in the western Indian Ocean

3.1.1 Vicariance versus dispersal

Oceanic dispersal played a central role in the early biogeography of Darwin and Wallace (Darwin 1859; Wallace 1876; Clark 1994; Hommersand, Fredericq, and Freshwater 1994), and for many decades afterwards. However, after continental drift and plate tectonics (Wegener 1912, 1966) became generally accepted, the field of biogeography shifted towards favouring vicariance explanations of biogeographic patterns, sometimes with dispersal serving as the 'poor cousin' explanation to be used only as a last resort when patterns were inconsistent with vicariance (de Queiroz 2005; Waters 2008; Nelson and Platnick 1981). Nelson (1979) famously characterized dispersalism as 'a science of the improbable, the rare, the mysterious, and the miraculous' because, at the time and before the molecular revolution in phylogenetics, inferring dispersal histories of organisms was too speculative. Thus many terrestrial and freshwater lineages occurring on Madagascar and some other islands in the Indian Ocean were long thought to have been present already before Madagascar separated from India and Africa during plate tectonic history (Yoder and Nowak 2006; Ali and Aitchison 2008). Support for vicariance hypotheses came both from morphology-based phylogenetic patterns that were consistent with the origin of Malagasy organisms from Africa or India, the two continents it was in connection with most recently, and from intuitive arguments based on the perceived limited dispersal ability of many of the highlighted lineages. For example chameleons, other lizards, frogs, lemurs, tenrecs, carnivores, and freshwater fish in Madagascar all were at one point or another thought to be examples of a common vicariance pattern (Yoder and Nowak 2006). However, a crucial component of vicariance hypotheses is the concurrent timing of geographic breakup of landmasses, and speciation events. Even when vicariance explanations are plausible based on phylogenetic relationships among species, and presumed dispersal abilities of the taxa in question, they must be rejected if the timing of speciation events is inconsistent with the geological history (Grande 1985; Lundberg 1993; Hunn and Upchurch 2001; Yoder and Nowak 2006). Molecular data offered a revolution in biogeography, as in phylogenetics in eneral, as they allow simultaneous estimation of relationships among species, and timing f speciation events (Drummond et al. 2006; Sanderson 2002). As massive DNA evidence is ccumulating, dispersal biogeography is currently enjoying a renaissance, with these new lata pointing towards relatively recent origins of lineages and thus more frequently than ot rejecting vicariance in favour of dispersal hypotheses for a wide range of taxa and andmasses (de Queiroz 2005; Yoder and Nowak 2006). Presumably, a common pattern is he occurrence of long distance dispersal events frequently enough to allow island olonization, but sufficiently rarely such that colonization events restrict gene flow and ventually lead to speciation (Gillespie et al. 2012).

he terrestrial and freshwater biota of the western Indian Ocean is no different – a review of ecent literature (Table 1, see also Yoder and Nowak (2006)) provides compelling evidence or the origin of the majority of the Indian Ocean island's biota at a much more recent eological time than the major tectonic events of the ancient Gondwana; the origin of these ;roups, then, has to be explained via Cenozoic dispersal rather than via Gondwanan icariance (Vences et al. 2001; Yoder and Nowak 2006; Kuntner and Agnarsson 2011a-b; Agnarsson and Kuntner In Prep). The ancestor of most lizards, frogs, lemurs, tenrecs, and at east some freshwater fish, are now all thought to have arrived via relatively recent long listance dispersal during the time the landmasses were in, or close to today's position Vences et al. 2004; Raxworthy, Forstner, and Nussbaum 2002; Austin, Arnold, and Jones .004; Rocha, Carretero, and Harris 2005; Rocha et al. 2006, 2007). In fact, our review (Table 1) lemonstrates that the majority, some 230 independent biogeographic histories (clades), nust be explained by ancestral dispersal followed by diversification, while only 16 clades eviewed are potentially old enough to be explained by diversification through Gondwanan icariance. One should bear in mind 1) that dated molecular phylogenies are far from nfallible. Calibration points are crucial, and fossils and geological events that are used for alibration are often not fully understood, and in any case usually only offer minimal ages. 'urther advances in molecular dating may yet favour vicariance for some of the clades ecently suggested to be examples of dispersal biogeography. Further, 2) the possibility xists that oceanic ridges, which may have facilitated dispersal, may have emerged above he surface periodically (Masters, de Wit, and Asher 2006). Thus, for example, not all Africa-Madagascar sister clades that are younger than the split between these landmasses must ave dispersed over ocean. However, recent paleo-oceanographic models predict ocean urrents favourable for colonization of Madagascar from Africa occurring exactly during the pisode when many taxa with presumably poor dispersal abilities, such as non-volant nammals, arrived (Ali and Huber 2010). The agreement between dated phylogenetic trees or multiple groups and paleogeographic modelling, and limited evidence for and lack of fit f data to land-bridge theories argue for transoceanic dispersal having been the major >iogeographical force in the region (Ali and Aitchison 2009; Ali and Krause 2011).

he sources of colonizers of western Indian Ocean islands are diverse, but Africa is dominant, specially for poorer dispersers, while elements from Asia– most notably India and Australasia also occur (Warren et al. 2010) among the better dispersers such as flying insects dipterans and dragonflies), fruit bats, and several groups of birds such as parrots and >asserines (Table 1). Madagascar, in turn, is the most common source of colonizers of the Mascarenes and Comoros, while the Seychelles have received dispersers from Madagascar, >ut also directly from Africa (Table 1). The complex tectonic history of the region (Ali and Aitchison, 2008) and many potential source landmasses mean that identifying the primary biogeographical forces contributing to the biota is challenging, and patterns are often taxor specific (Table 1). The modes of dispersal must be either aerial (wind or bird-assisted; Gillespie et al. 2012), ocean rafting over the more than 400 km wide Mozambique Channel between Africa and Madagascar, or from even further afar, or a mix of both (Table 1), although at mentioned above, potential ridges above the ocean may have facilitated island hopping in the past. In some cases, recent colonisations have been assisted by human transport (Vences et al 2004). Whatever the route and means of dispersal, the available data clearly imply the importance of dispersalism. Thus, long distance dispersal may well be the study of the improbable and the rare but hardly the mysterious nor the miraculous (see Nathan 2006). As it turns out, rare and individually improbable events are extremely important historically, and indeed such 'waif biota' (Carlquist 1966) is responsible for much of the distribution and diversification of organisms on islands.

Nevertheless, vicariance remains an important biogeographical force in the region and molecular data has corroborated vicariance hypotheses for a number of extant groups (see below, Table 1, and Yoder and Nowak (2006). Many extinct Malagasy groups are also though to represent vicariant elements. Indeed the relative scarcity of extant vicariant groups may relate to faunal turnover following meteorite impact and associated volcanic activity 65 Ma (see below). It is also worth noting that detecting vicariance may be challenging due to sampling errors, extinction, and species expansion following disappearance of historica geological barriers (Upchurch 2008). Current methods may thus to some extent be underestimating the importance of vicariance. Calibrating species phylogenies using island ages, commonly done when fossil data are not available, including in our own recent studies (Kuntner and Agnarsson 2011a-b), may also be misleading (Heads 2011). Clades endemic or islands may be older than the islands, having e.g. persisted on older nearby islands that are now submerged, and thus using island age to calibrate phylogenies will tend to bias studies towards accepting dispersal hypotheses over vicariance. Future studies, through improved knowledge of the fossil records and better understanding of 'molecular clocks', will no doubt lead to reinterpretation of some dispersal hypotheses. Nevertheless, overwhelming evidence suggests that rather than treating dispersalism as explanation of the last resort, transoceanic dispersal has strongly influenced the distribution of organisms globally (Carlquist 1966; Crisp et al. 2009; Gillespie et al. 2012). In sum, ancestors of most of the biota of Madagascar, and indeed other Wallacean islands in the western Indian Ocean, arrived there in the last 65 Ma, after it was isolated from other landmasses, and the current evidence does not favour tentative Cenozoic land-bridges as influential in the colonization of Madagascar or other islands (Ali and Aitchison 2009; Ali and Krause 2011).

3.1.2 Cenozoic dispersal from Africa to Madagascar, and from Madagascar to the Comoros, Seychelles, and Mascarenes – The dominant patterns

The Cenozoic dispersal from Africa model (i.e. younger than 65.5 Ma) seems to apply to the majority of the Malagasy fauna and flora (Yoder and Nowak 2006), although Warren et al. (2010) conclude that Asian elements are nearly as important as African ones in Madagascar, especially among vertebrates. The colonization of Madagascar is often followed by subsequent colonization of the Seychelles, Comoros and Aldabra to the north, and the chain of volcanic Mascarene islands (Réunion, Mauritius and Rodrigues) to the south and east (Fuller, Schwarz, and Tierney 2005; Yoder and Nowak 2006; Raxworthy et al. 2007; Le Pechon et al. 2010). Examples of Cenozoic dispersal from Africa to Madagascar include arious elements of herpetofauna such as geckos, tortoises, scincid lizards, and colubrid nakes (Raxworthy, Forstner, and Nussbaum 2002; Vences et al. 2001; Hume 2007; Nagy et l. 2003; Austin, Arnold, and Jones 2004; Caccone et al. 1999; Mausfeld et al. 2000), mammals Tattersall 2006; Masters, de Wit, and Asher 2006; Masters, Lovegrove, and de Wit 2007; tussell, Goodman, and Cox 2008; Goodman, Buccas, et al. 2010), fig wasps (Kerdelhue, Le Clainche, and Rasplus 1999), allodapine bees (Fuller, Schwarz, and Tierney 2005), carabaeinae dung beetles (Sole and Scholtz 2010), carpenter bees (Rehan et al. 2010), pierid utterflies (Nazari et al. 2011), fig wasps (Kerdelhue, Le Clainche, and Rasplus 1999), spiders Agnarsson and Kuntner 2005; Agnarsson et al. 2010; Kuntner and Agnarsson 2011a-b), ants Fisher 2007), freshwater crabs (Daniels et al. 2006; Klaus, Schubart, and Brandis 2006), snails Kohler and Glaubrecht 2010), parasitic flatworms (Verneau, Du Preez, and Badets 2009; 'erneau et al. 2009) and many groups of plants (Renner 2004; Weeks, Daly, and Simpson .005; Dick, Abdul-Salim, and Bermingham 2003; Kita and Kato 2004; Bartish et al. 2011; ichaefer, Heibl, and Renner 2009; Manns and Anderberg 2011; Pokorny, Olivan, and Shaw :011; Renner 2004), although many plants seem to have a more complicated biogeographical ustory (Renner 2004). This pattern of colonization is also seen intraspecifically, for example he bryophyte *Calyptrochaeta asplenioides* colonized Madagascar from E. Africa about 6 Ma, hen subsequently the Mascarenes about 3 Ma from Madagascar, also likely reaching the Comoros from Madagascar (Pokorny, Olivan, and Shaw 2011). Many groups of birds have lso arrived in Madagascar via Cenozoic dispersal from Africa (Yamagishi et al. 2001; Marks nd Willard 2005; Warren et al. 2003, 2005; Groombridge et al. 2002), but being generally xcellent dispersers, many have also arrived from other continents further away (see below). n several of these groups, such as various groups of bats (Weyeneth et al. 2008, 2011 5oodman, Chan, et al. 2010), skinks (Austin, Arnold, and Jones 2009), day geckos (Harmon t al. 2008; Raxworthy et al. 2007), nephilid spiders (Kuntner and Agnarsson 2011a-b), fig vasps (Kerdelhue, Le Clainche, and Rasplus 1999), and in plants (Le Pechon et al. 2009, !010; Wikstrom et al. 2010; Pokorny, Olivan, and Shaw 2011), colonization of Madagascar vas likely followed by dispersal from Madagascar to other islands, including the Comoros nd/or the Mascarenes. In at least some groups the data are consistent with stepwise olonization to Mauritius and Réunion, and on to Rodrigues (Venkatasamy et al. 2006; Kuntner and Agnarsson 2011b). In fact, Mauritius emerges as an important source of olonizers to Réunion (Fig. 2), consistent with its older age. In coastal lizards Madagascar vas colonized from Australasia, followed by colonization of Comoros and Mauritius from Madagascar (Rocha et al. 2006). There are, however, also cases of direct colonization of Mascarenes or Comoros from Africa, for example in birds (Warren et al. 2006), land snails Rowson, Tattersfield, and Symondson 2011), and plants (Anthony et al. 2010). Similarly, East Africa is an important source of colonizers of the Seychelles, as seen in *Archaius* :hameleons (Townsend et al. 2011), freshwater crabs (Daniels 2011) and land snails Rowson, Tattersfield, and Symondson 2011).

3.1.3 Diverse origin and biogeographic patterns of excellent dispersers

itretches of ocean present the least effective barriers to strong flyers such as birds, bats, bees, ind dragonflies, and to plants that disperse with birds or with readily airborne seeds - and :hese groups have the most diverse regions of origin in the Indian Ocean. Indeed, many of he 'secondary' colonization routes for Madagascar depicted with grey arrows in Fig. 2 have oredominantly been taken by excellent dispersers, while intermediate-poor dispersers tend :o follow similar colonization patterns, based on trade winds, currents, and other abiotic

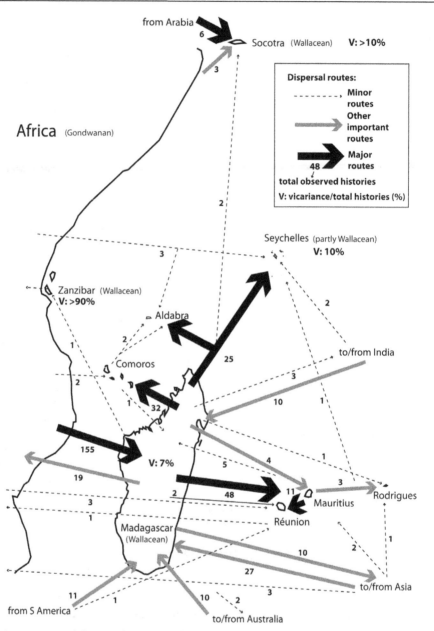

Fig. 2. A summary of dispersal events and vicariance elements on the islands of the western Indian Ocean based on our review of literature (Table 1), and that of Yoder and Nowak (2006). Major routes are indicated by thick black arrows and represent our interpretation of the patterns, with the total number of documented events in our review shown by number next to the arrow. Note that patterns shown are based on relatively recently published papers and taxa and histories from older literature may change interpretations somewhat.

For example, colonization of Comoros is probably more complex than shown here, the importance of Africa may be more as several bird species not included here have colonized Comoros from Africa. The historical importance of Asia as source of colonizers for Africa may also be greater than implied here (see Warren et al. 2010). Also, certain rare routes are omitted for simplicity - there is at least one possible colonization of Comoros from India, by scops owls (Fuchs et al. 2008).

factors. For example, oscine passerines arrived via overwater dispersal from Australia to Asia, moving over the Indian Ocean and eventually to Africa, where they radiated (Jonsson and Fjeldsa 2006), parrots probably colonized the Mascarenes from India (Hume 2007), scops owls colonized Asia from the Indian Ocean (Fuchs et al. 2008), and magpie robins colonized Madagascar from Asia (Sheldon et al. 2009). Among flying mammals, *Pteropus* bats repeatedly colonized the islands from Asia (O'Brien et al. 2009), and *Triaenops* bats colonized Madagascar repeatedly from Africa resulting in several Malagasy radiations (Russell, Goodman, and Cox 2008). For *Rousettus* bats current phylogenies are inconclusive, but they may have arrived from Asia, or the Middle East (Goodman, Chan, et al. 2010). Flying insects have also arrived from all directions, e.g. *Zaprionus* dipterans colonized the Indian Ocean islands enroute from Asia to Africa (Yassin et al. 2008). Similarly, dragonflies also dispersed over ocean to colonize the western Indian Ocean islands from India, Africa, and Asia (Dijkstra 2007). One species dispersed from Asia to Mauritius (Clausnitzer 2003), while a damselfly colonized Pemba island most likely from Madagascar (Dijkstra, Clausnitzer, and Martens 2007). Some dragonfly species are widespread, for example throughout the Seychelles, Africa and Asia, and are known to migrate long distances over water (Clausnitzer 2003). Allodapine bees colonized Madagascar several times from Africa in the Miocene, and then moved on to Asia and Australia (Fuller, Schwarz, and Tierney 2005). Several plant groups also show a pattern of Australasian origin of Indian Ocean islands colonizers (Kulju et al. 2007; Renner et al. 2010), with others arriving from S. America to Madagascar (Bone et al. 2011) or directly to the Mascarenes (Cuenca, Asmussen-Lange, and Borchsenius 2008). Among non-flyers, the pattern is more difficult to explain in reptiles, where e.g. coastal lizards as potentially poor dispersers nevertheless achieved transoceanic dispersal from Australia or Indonesia to Madagascar and from there to Africa, Comoros and the Mascarenes (Rocha et al. 2006). However, many factors indicate that these lizards are exceptionally successful at over ocean dispersal, such as their presence on very numerous islands and islets throughout the Pacific and western Indian Ocean, and the low genetic divergences found between species in the western Indian Ocean and the Pacific ocean, and among western Indian Ocean islands. Finally, *Pristionchus pacificus* nematodes that are cosmopolitan parasites of scarabid beetles have arrived to Réunion island from many different directions (Herrmann et al. 2010), demonstrating the ability of this nematode to travel with various different species and thus collectively be an excellent disperser.

3.1.4 The impact of 'reverse colonization' in the Indian Ocean

The traditional thinking in biogeography has long been to view islands as sinks and continents as sources of colonizers. However, results from many recent studies highlight over ocean 'reverse colonization' events, from islands to continents as an important biogeographical force (Sturge et al. 2009; Heaney 2007; Bellemain and Ricklefs 2008) especially from large islands close to continents (Gillespie et al. 2012). Reverse colonization has played an important role in the Indian Ocean, including islands as sources for other islands as outlined above, a pattern often seen in organisms that use birds as dispersal vectors (Gillespie et al. 2012) (Fig. 2). Chameleons, as an example, originated and radiated in Madagascar after the Gondwana breakup, then subsequently colonized Africa via rafting from Madagascar (Raxworthy, Forstner, and Nussbaum 2002). They also reached the Comoros, Seychelles and Réunion, resulting in minor radiations. *Phelsuma* geckos are also thought to have originated in Madagascar, though their ancestor may have arrived there via dispersal from Africa, where its sister lineage occurs (Austin, Arnold, and Jones 2004). *Phelsuma* subsequently dispersed overwater multiple times colonizing the Mascarenes, Aldabra, Comoros, Seychelles, Andamans and Pemba. Coastal lizards (*Cryptoblepharus*) colonized Madagascar from Australia or Indonesia, then diversified in Madagascar, and subsequently 'reverse colonized' the East African coast, the Comoros islands and Mauritius (Rocha et al. 2006). In fact, the colonization of mainland Africa may have been a result of multiple island hopping events; African haplotypes nest within a Zanzibar clade (Rocha et al. 2006) consistent with the colonization of Zanzibar and the subsequent colonization of Africa from these offshore islands. A platycnemis damselfly is thought to have colonized Pemba from Madagascar, but not moved to the mainland due to unsuitable habitat (Dijkstra, Clausnitzer, and Martens 2007). Scops owls reverse colonized Asia from western Indian Ocean islands (Fuchs et al. 2008). *Zaprionus* dipterans reached Africa from Asia where the Indian Ocean islands served as stepping stones (Yassin et al. 2008). Similarly, though in the opposite direction, Gecarcinucoidea freshwater crabs reached Asia from Africa via the Indian Ocean islands (Klaus, Schubart, and Brandis 2006), and allodapine bees dispersed from Africa to Australia, also via the islands (Schwarz et al. 2006). One of the most spectacular examples of Madagascar as a source of colonizers comes from the coffee family, Rubiaceae (Wikstrom et al. 2010). Plants of this family have dispersed 'out of Madagascar' at least 13 times, colonizing Africa at least three times, Asia at least twice, Seychelles twice or more, Comoros multiple times, and the Mascarenes at least twice (Wikstrom et al. 2010; Maurin et al. 2007).

3.1.5 Vicariant elements in the Indian Ocean and the great faunal turnover

Despite the renaissance of dispersalism, vicariance remains crucially important in biogeography, and for the western Indian Ocean, vicariance explanations are supported by molecular data for several lineages. Lineages that seem to represent ancient Gondwanan Malagasy radiations include: typhlopid blindsnakes (Vidal et al. 2010), boid snakes, podocnemid turtles, and iguanid lizards (Noonan and Chippindale 2006), ranid frogs (Bossuyt et al. 2006; Bossuyt and Milinkovitch 2000), the extinct elephant birds (Cooper et al. 2001), and possibly some cichlid and rainbow fishes (Chakrabarty 2004; Sparks 2004; Sparks and Schelly 2011; but, see Ali and Aitchison (2008)). Many extinct vertebrate lineages with a fossil record in Madagascar from ca. 88-65 Ma, are also thought to be vicariant (Masters, de Wit, and Asher 2006). These include several dinosaurs, 'proto-birds' seemingly related to *Archaeopteryx*, and gondwanatherian mammals (see Masters, de Wit, and Asher (2006) for summary and references). On the granite Seychelles, vicariance seems the best explanation for the existence of ancient caecilian amphibians (Zhang and Wake 2009) whose closest relatives are from India (Gower et al. 2011) and sooglossid frogs (Biju and Bossuyt 2003), oth lineages of burrowing organisms. On Socotra, vicariance may be more important than n the other Wallacean islands. Putative examples include the Socotran chameleon (Macey t al. 2008), geckos (Arnold 2009; Gamble et al. 2008), and possibly some freshwater crabs Shih, Yeo, and Ng 2009). One reason to expect greater importance of vicariance in Socotra is hat it separated from the continental landmasses long after the catastrophic meteor impact nd Deccan volcanism, unlike Madagascar (see below), and it did not become mostly ubmerged in the sea during high ocean levels, unlike the Seychelles. Thus our prediction is hat a greater portion of as yet studied lineages will show vicariance patterns on Socotra ompared to Madagascar or Seychelles.

ome invertebrates may also be of vicariant origin on the islands of the western Indian Ocean, uch as crayfish (Toon et al. 2010), giant pill-millipedes (Wesener, Raupach, and Sierwald 2010; Vesener and VandenSpiegel 2009), migid and archaeid spiders (Griswold and Ledford 2001; Vood, 2008; Wood, Griswold, and Spicer 2007) and Troidini butterflies (Braby, Trueman, and astwood 2005). Several groups of plants are also thought to be vicariant 'relicts', evidenced iot necessarily by available molecular dates but by shapes of phylogenetic trees (Heads 2009): he Malagasy genus *Humbertia*, for example, is sister to the presumably old family Concolculaceae, and the genus *Takhtajania* is sister to Winteraceae (Schatz 1996, cited in Yoder ind Nowak (2006). Ferns (Lindsaeaceae) from Seychelles are most closely related to the ones rom Sri Lanka and India, and long branches indicate an ancient split of these groups Lehtonen et al. 2010). Similarly, *Didymeles* containing two species distributed in Madagascar ind the Comoros, is sister to the order Buxales. For this last split an approximate molecular late estimate is available and, at 99 Ma for the crown group Buxales, is consistent with icariance (Anderson, Bremer, and Friis 2005). However, several of these vicariance ypotheses still require testing through dated phylogenies. Recent examples, including from ur own work, show that speculations based on phylogenetic patterns, presumed poor lispersal abilities, and suspected old age of a group, such as clitaetrine spiders (Kuntner 2006), till fail to find support for vicariance in dated molecular phylogenies (Agnarsson and Kuntner n prep.). Upcoming dated phylogenies will also test the hypothesized Gondwanan origin on Madagascar of migid and archaeid spiders (Griswold and Ledford 2001; Wood, 2008; Wood, Griswold, and Spicer 2007), and other groups.

The relative scarcity of unequivocal vicariance examples may reflect the apparent massive aunal turnover of vertebrates, and likely also other groups, that is hypothesized to have esulted from a meteor impact some 65 Ma in western India (Krause (2003) cited in Yoder ind Nowak (2006)). The meteor possibly triggered the massive Deccan volcanism in the Vestern Ghats that may have spanned 30.000 years and resulted in the release of huge imounts of volcanic gases, with direct effects on biodiversity and indirect effects through a lrop in global temperature (Schulte et al. 2010; Bryan et al. 2010; Negi et al. 1993). While the vidence for faunal turnover stems mostly from vertebrates, in part due to their better fossil ecord, this is probably a more general pattern. It seems plausible that a mass extinction :vent, not restricted to vertebrates, occurred around 65 Ma, rendering the post-impact Madagascar, effectively, a 'Darwinian' or *de novo* island for many lineages. This hypothesis vould explain the dominant support for the Cenozoic dispersal scenario among a very liverse set of organisms in Madagascar, and the relative scarcity of vicariant elements. In urn, the partial submergence of the Seychelles during high ocean levels may have dramatically reduced vicariant elements, while for Socotra, vicariant elements should no have been wiped out by such catastrophic events.

A curious vicariance example is the presence of caecilians in Seychelles, and equally curiou is their absence from Madagascar. If a vicariance hypothesis is true, it would not b surprising to discover hitherto not found caecilian fossils in Madagascar. Such a find woul verify vicariance and the faunal turnover hypotheses.

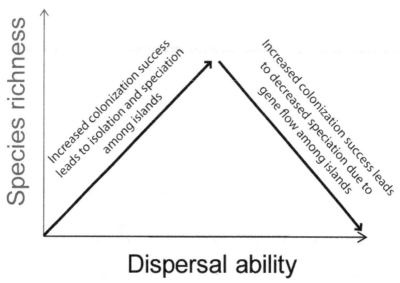

Fig. 3. A conceptual model of dispersal biogeography (adapted from Agnarsson and Kuntner In Prep). Dispersal ability positively relates to probability of island colonization. Initially increased dispersal ability leads to colonization followed by isolation and speciation among islands. As dispersal ability increases further, more islands are colonized but dispersal begins to retain gene flow among islands and thus restricting speciation. The highest species richness is thus expected in intermediate dispersers.

3.2 Biogeography and phylogeography of spiders in the Indian Ocean

The spider fauna of Madagascar and other Indian Ocean islands has been studied for a long time, but very few biogeographical studies exist. A few groups of spiders have been recently revised, or revisited, in some cases offering preliminary biogeographical speculations (Coyle 1995; Platnick 1995; Griswold 1997; Corronca 2003; Huber 2003; Agnarsson and Kuntner 2005; Agnarsson 2006; Knoflach and Van Harten 2006; Kuntner 2006, 2007; Huber and El-Hennawy 2007; Wood, Griswold, and Spicer 2007; Wood 2008; Logunov 2009; Lyle and Haddad 2010; Ubick and Griswold 2011). Though the evidence is not strong, these studies generally indicate the closest living relatives of the Indian Ocean island fauna being from Africa, consistent with the Cenozoic dispersal model outlined above, the likely exception being the seemingly Gondwanan migids and arhaeids (Griswold and Ledford 2001; Wood, 2008; Wood, Griswold, and Spicer 2007). Testing broad biogeographical hypotheses in a molecular phylogenetic framework has just begun, with our own studies on three nephilid

Clitaetra, Nephila, Nephilengys) and four theridiid lineages (*Anelosimus, Argyrodes, Faiditus, Neospintharus*) (Agnarsson and Kuntner 2005; Agnarsson et al. 2010; Kuntner and Agnarsson 011a-b; Agnarsson and Kuntner In Prep).

he social theridiid spiders of the genus *Anelosimus* were once thought to only occupy the neotropics (with solitary congeners found in Europe), however, we have recently discovered radiation of subsocial *Anelosimus* in the mountains of eastern and northern Madagascar Agnarsson and Kuntner 2005). Though existing phylogenies are ambiguous as to what the losest relatives of this Malagasy radiation are, the most recent study suggests the sister ineage is African, and the group as a whole much too recent to have a vicariant history in he region (Agnarsson et al. 2010). Thus, dispersal from Africa seems the most likely source of social Malagasy *Anelosimus*. Curiously, the solitary *A. decaryi*, found on the beachfront in Madagascar, Comoros and Aldabra atoll, represents a different evolutionary lineage, having ndependently colonized the western Indian Ocean (Agnarsson et al. 2010), and this olonization must be very recent given the age of Aldabra and Comoros. Apparently gene low is retained among the coastal *A. decaryi* between Madagascar and Mayotte, however vithin Mayotte the lineage has diverged into two species including the non-coastal *A. melie*. It is remarkable that the 'solitary' beachfront habitat was colonized from overseas, ather than by the *Anelosimus* lineage that lives in the nearby mountains, suggesting that hese spiders more readily disperse between landmasses than switch habitats. The closest elatives of *A. decaryi*, and its sister species *A. amelie* from Mayotte, are unknown; they nest vithin an American clade, also containing European species, in the study of Agnarsson et al. 2010), but *Anelosimus* spiders are so poorly documented in Africa that biogeographical nterpretations are inconclusive at this stage.

Vithin the spider family Nephilidae, three genera are distributed with one to several species n the islands of the western Indian Ocean (Fig. 4). The best known is *Nephila*, which was hought to have two species in the region, the widely spread *N. inaurata* found in Africa, Madagascar, and numerous of the smaller islands in the region, and *N. ardentipes* thought to be endemic to Rodrigues (Kuntner, Coddington, and Hormiga 2008). Being that *Nephila* occupies many landmasses globally and that most species show extremely wide ranges that ometimes spread over several continents (Kuntner and Agnarsson 2011b), *Nephila* is hought to be an excellent aerial disperser. Its excellent dispersal ability begged the question f the Rodrigues *A. ardentipes* might not simply represent a population of the widespread *N. naurata*. We set out to test that hypothesis and our phylogeographic analysis based on the ampling of *Nephila* over the island area from Mayotte through Madagascar, Réunion, Mauritius and Rodrigues revealed that *Nephila* likely maintains, or recently maintained, gene flow over the islands, preventing speciation such that all these populations indeed belong to the widespread *N. inaurata* (Kuntner and Agnarsson 2011b). The phylogeny and haplotype network analyses both suggested an ancestral dispersal from Africa to Madagascar and Comoros (likely, also to the Seychelles, though we did not have specimens o test it), and from Madagascar to the Mascarenes, possibly stepwise to Mauritius and Réunion, and then on to Rodrigues, all these taking place relatively recently, less than 4 Ma Kuntner and Agnarsson 2011b). Yet, our results were consistent with a recent cessation of gene flow, with each of the Mascarene islands and Madagascar having accumulated some sland-unique haplotypes, a scenario we referred to as a possible 'speciation in process'.

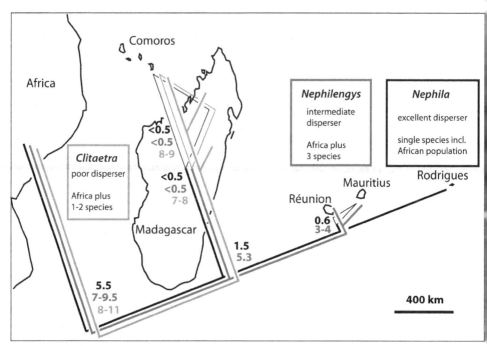

Fig. 4. Distribution, colonization patterns, and mtDNA genetic structure and divergence among three codistributed and closely related lineages of Nephilidae spiders, differing in dispersal ability. The best disperser, *Nephila*, is most widely distributed, but shows the shallowest divergences among islands, with a single species in the area. The poorest disperser, *Clitaetra*, is the most narrowly distributed, and with the deepest genetic divergences in shared clades, and additional genetic structure lacking in the other species (within Madagascar, between Madagascar and Mayotte). It is represented by 1-2 species in the region. The intermediate disperser, *Nephilengys*, has colonized most islands, but appears to do so sufficiently rarely such that island colonization is followed by speciation. Thus, there is an endemic species each on Madagascar, Réunion, and Mauritius, plus African species. Data from Kuntner and Agnarsson (2011a-b); Agnarsson and Kuntner (in prep). See text for details.

The second nephilid genus, *Nephilengys*, had been thought to be only represented in the region by a single widespread species *N. borbonica*, which is distinct from the related African mainland *N. cruentata* (Kuntner 2007). However, given that *Nephilengys* is much more narrowly distributed than *Nephila*, and is thought to be a poorer disperser, this scenario also seemed suspect. We set out to study the phylogeography of this species across the same islands as above, and found that indeed, each of the colonized Mascarene islands, Mauritius and Réunion had a genetically distinct population. We thus delimited all island populations into three species: *N. livida* inhabiting Madagascar and Mayotte (and possibly, other Comoros and Seychelles, though that remains to be tested), *N. borbonica* endemic to Réunion, and *N. dodo* endemic to Mauritius (Kuntner and Agnarsson 2011a). Consistent with its lesser dispersal ability, no *Nephilengys* was found in Rodrigues. However, the colonization pattern in *Nephilengys* was otherwise identical to *Nephila*, with relatively recent colonization of Madagascar and Comoros, and subsequent colonization of the Mascarenes.

In contrast, the third nephilid genus, *Clitaetra*, has not colonized the Mascarenes nor Seychelles, but is confined to Africa, Madagascar, and the Comoros, with an additional species in Sri Lanka (Kuntner 2006). The very narrow distribution of the genus, and its specialized arboricolous habits with seemingly limited local reach (Kuntner and Agnarsson 2009) hinted at poor dispersal abilities of *Clitaetra* spiders in comparison to its close relatives *Nephila* and *Nephilengys*. We should emphasize here that dispersal abilitities of nephilid spiders were assessed a priori and not from the resulting genetic divergencies, thus avoiding circularity (Agnarsson and Kuntner in prep.). Our phylogeographic analyses of the Indian Ocean *Clitaetra* populations, including representatives of all but one African species of the genus, confirmed the prediction that the poor disperser will exhibit the most limited distribution and at the same time the greatest genetic structure among islands.

Preliminary data on Argyrodinae spiders also hint at recent African origin of the species found in Madagascar, Comoros, Mauritius and Réunion, but remarkably suggest that the native Rodrigues *Argyrodes*, the only species we encountered on the island, belongs to a different genetic lineage than those on other islands, although the affinities of the Rodrigues species are as yet unknown (Agnarsson et al. unpublished)

In sum, spiders in general, and certainly the lineages we have examined so far are mostly consistent with the Cenozoic dispersal model, with the colonization of Madagascar from Africa, and the subsequent colonization of Comoros, and the Mascarenes from Madagascar (Fig. 2). As noted above, the likely exceptions are the Gondwanan elements on Madagascar of the families Migidae and Archaeidae. Dated phylogenetic studies indicate colonization of Madagascar from Africa 2-8 Ma, and subsequent colonization of Comoros and the Mascarenes, most likely from Madagascar no more than 4 Ma. The patterns observed are clearly dependent on dispersal abilities of the taxa. In all cases we found a distinct and well supported western Indian Ocean clade with a relatively deep divergence between it and the sister African clade. However, the better the disperser, the less genetic divergence it shows between Africa and the outer Indian Ocean islands. All taxa we examined have reached the Comoros, but only the poorest disperser (*Clitaetra*) shows divergences between Madagascar and the Comoros. Only the better dispersers among the lineages examined have reached the Mascarenes, and only the best dispersers (*Nephila, Argyrodes*) reached the isolated island of Rodrigues, and only the intermediate disperser *Nephilengys* has speciated within the Mascarenes.

3.3 The role of dispersal ability in distribution and diversification

Our work on the Indian Ocean nephilids was inspired by the conceptual model presented in Agnarsson and Kuntner (In prep, see Fig. 3). The model predicts an approximately linear relationship between a taxon's dispersal ability and the number of islands it occupies, but that due to the interplay of colonization and gene flow, intermediate dispersers, rather than good or poor dispersers, should be most diverse on archipelagos. One should note that this relationship can be complicated if there is a change in dispersal abilities upon colonization and radiation. Many island taxa have secondarily lost dispersal ability after colonizing isolated islands, including many groups of spiders (Gillespie et al. 2012). Nephilids offered an ideal opportunity for initial test of this model, with three closely related lineages occurring in the region, differing in dispersal abilities. Taken together, our studies on these related and codistributed spider lineages, outlined above, are consistent with the model: *Nephila* as the best disperser is the most widespread, *Clitaetra* as the poorest disperser the most narrowly distributed and most genetically structured, and *Nephilengys* as the intermediate disperser the most diverse (Fig. 4).

Is this pattern repeated in the bigger picture examination of the range of taxa examined here? In other words, how does dispersal ability in general relate to diversity across archipelagos? Table 1, and table 1 from Yoder and Nowak (2006), include a large range of organisms differing from poor to excellent dispersers. Of the taxa examined the least diverse in the region are the very poor, and the extremely good, transoceanic dispersers. Active and strong day-fliers are arguably the best dispersers, including birds, fruit bats, and dragonflies. Many plants are also excellent dispersers via their airborne seeds, as are some spiders that get readily airborne using silk strands. Oceanic birds, probably the best dispersers of all, have colonized practically all vegetated islands in the western Indian Ocean , but have not radiated extensively (Yoder and Nowak 2006; Hume 2007). Among these, for example, the scooty tern (*Onychoprion fuscatus*) occurs on a large number of islands, pantropically in all major oceans, with clear evidence of gene flow among islands and even oceans that has prevented diversification, though several 'clinal' subspecies are recognized. In fact, the genus *Onychoprion* though extremely widespread and abundant, contains only four species in total. *Triaenops* fruit bats, also good dispersers, have repeatedly colonized Madagascar and adjacent islands in the relatively recent past, but remain species poor (Russell, Goodman, and Cox 2008), presumably due to continuous gene flow across their Malagasy range. Some of the strongest flying insects such as dragonflies have colonized all the islands but only diversified very moderately and many species are widespread and known to disperse long distances overwater (Dijkstra 2007). Several species that stick to strong flying organisms, such as many plant seeds and invertebrates that stick to birds, can also be excellent dispersers, and certain rafters such as coconuts are also excellent dispersers (e.g. Gillespie et al. 2012). Night flying organisms, such as most bats and moths are in general less capable of dispersal over ocean. Yet these are found on practically all the major islands and some bats have endemic species on many of the larger and older Darwinian islands. Poorer dispersers still are many of those that do not fly and rely on rafting to disperse. Even these have reached a large number of islands, and typically show patterns of single-island endemism. Exceptionally diverse among these are chameleons on Madagascar, *Phelsuma* geckos, and coastal lizards, conforming to the model prediction that diversity on large Wallacean islands will be skewed towards relatively poor dispersers (Fig. 3). Non-volant mammals and amphibians, are even poorer dispersers, mammals being relatively large and thus requiring much larger 'rafts' than most reptiles and amphibians or insects, and amphibians are probably particularly unlikely to survive a trip on an oceanic raft due to their general sensitivity to dessication. These groups are absent on nearly all the islands, except Madagascar where they have radiated extensively (Yoder and Nowak 2006), typically as a result of a single successful colonization event. This highlights the extreme rarity of such events and thus how poor dispersal ability can severely restrict diversification among islands, but may promote it within the few islands they by chance manage to colonize (Fig 3). Indeed, the western Indian Ocean has been colonized only by a small subset of the lineages 'available' on nearby continents, and the lineages lacking are mostly poor dispersers such as wingless insects, burrowing animals, and several orders of non-volant mammals.

The excellent dispersers are relatively species poor due to maintenance of gene flow. Many oceanic birds have populations on numerous islands, among whom gene flow is maintained. Similarly, several dragonfly species are widespread in the region, with among island speciation prevented by gene flow. The exceptions from this pattern are poor dispersers that are present on Wallacean islands, some of which are very diverse. These include vicariant lineages that thus did not disperse over ocean to reach the island, and taxa that have reached the Wallacean islands via dispersal, such as carnivores, tenrecs and rodents. In both cases, their relatively poor dispersal ability means that isolation among populations is created even by minor barriers such as rivers, mountain ridges, different vegetation types etc, which thus may lead to diversification. Here, again, dispersal ability relates to diversification patterns. However, these poor dispersers have only reached Madagascar, which invites the question of why it is that poor dispersers are found only on Wallacean islands, even when they had to disperse to get there. The general answer is likely simple probability, Wallacean islands have, in general, been around much longer than volcanic Darwinian islands, thus there has been a much longer time for that rare dispersal event to occur. Wallacean islands also started their lives close to continents, gradually moving away. Finally, in the case of Madagascar, not only has it been around for a long time, it is also a very large island and is thus a huge target making colonization there, by any disperser, much more likely than on smaller islands.

4. Conclusions

We summarized recent data on the biogeography of the terrestrial and freshwater fauna and flora of the western Indian Ocean, with some emphasis on our own work on spiders. As convincingly demonstrated by Yoder and Nowak (2006) for Madagascar, the dominant biogeographical pattern for the western Indian Ocean is ancestral Cenozoic dispersal from Africa to Madagascar. We show here that dispersal from Madagascar, in turn, to the smaller islands to the north and east of Madagascar is the main mode of colonization of most major islands in the western Indian Ocean. Two other notable elements are, first, the diverse origin of the best dispersers, having arrived from Africa, Asia, Australia, and the Americas, clearly indicating the importance of considering different dispersal abilities of lineages when studying broad biogeographical patterns. And second, the presence of a limited number of extant vicariant (Gondwanan) lineages. We believe that the scarcity of vicariant lineages, and the dominant pattern of colonization of Madagascar from the physically most proximate current-day continent are best explained with a mass extinction event occurring approximately 65 Ma having wiped out much of the Malagasy biota. Hence despite being geographically a Wallacean island that once formed a part of Gondwana, from a biogeographical perspective, Madagascar by and large shows patterns expected from a large Darwinian island. We also conclude that dispersal ability of taxa not only affects their distribution but also patterns of genetic divergences and speciation, such that species richness peaks at intermediate dispersal abilities.

5. Acknowledgments

Primary funding for this work came from the National Science Foundation (grant DEB-1050187-1050253 to I. Agnarsson and G. Binford). Additional funding came from the European Community 6th Framework Programme (a Marie Curie International

Reintegration Grant MIRG-CT-2005 036536 to M. Kuntner), and the National Geographic Society (grant 8655-09 to the authors), and this is contribution number 9 resulting from the 2008 Indian Ocean expedition, funded by the Slovenian Research Agency (grant Z1-9799 0618- 07 to I. Agnarsson) and the National Science Foundation (grant DEB-0516038 to T Blackledge). Greta Binford, Erin Saupe, Rosemary Gillespie and Lauren Esposito contributed in various ways to this work through collaborations and by sharing ideas Special thanks to Rosemary Gillespie, Christopher Raxworthy, and Kesara Anamthawat-Jónsson for comments that greatly improved the manuscript. Clarifications for the distribution and diversity of various lineages were kindly provided by Hanno Schaefer, Ben Rowson, Nikals Wikström, and James Ackerman.

6. References

Agnarsson, I, and M Kuntner. In Prep. A conceptual model of dispersal biogeography predicts highest diversity of intermediate dispersers.

Agnarsson, I. 2006. Asymmetric female genitalia and other remarkable morphology in a new genus of cobweb spiders (Theridiidae, Araneae) from Madagascar. *Biological Journa of the Linnean Society* 87 (2):211-232.

Agnarsson, I., and M. Kuntner. 2005. Madagascar: an unexpected hotspot of socia *Anelosimus* spider diversity (Araneae : Theridiidae). *Systematic Entomology* 30 (4):575-592.

Agnarsson, I., M. Kuntner, J.A. Coddington, and T.A. Blackledge. 2010. Shifting continents not behaviours: independent colonization of solitary and subsocial *Anelosimus* spider lineages on Madagascar (Araneae, Theridiidae). *Zoologica Scripta* 39:75-87.

Ali, J. R., and J. C. Aitchison. 2008. Gondwana to Asia: Plate tectonics, paleogeography and the biological connectivity of the Indian sub-continent from the Middle Jurassic through latest Eocene (166-35 Ma). *Earth-Science Reviews* 88 (3-4):145-166.

Ali, J. R., and J. C. Aitchison. 2009. Kerguelen Plateau and the Late Cretaceous southern-continent bioconnection hypothesis: tales from a topographical ocean. *Journal o Biogeography* 36: 1778–1784.

Ali, J. R., and M Huber. 2010. Mammalian biodiversity on Madagascar controlled by ocear currents. *Nature* 643: 653-657.

Ali, J. R., and D. W. Krause. 2011. Late Cretaceous bioconnections between Indo-Madagascar and Antarctica: refutation of the Gunnerus Ridge causeway hypothesis. *Journal o Biogeography* 38: 1855-1872

Anderson, C. L., K. Bremer, and E. M. Friis. 2005. Dating phylogenetically basal eudicots using rbcL sequences and multiple fossil reference points. *American Journal o Botany* 92 (10):1737-1748.

Anon. 1998. *Ile Rodrigues. Carte Géologique au 1:25 000. Schéma hydrogéologique.* Paris.: Geolab.

Anthony, F., L. E. C. Diniz, M. C. Combes, and P. Lashermes. 2010. Adaptive radiation in Coffea subgenus Coffea L. (Rubiaceae) in Africa and Madagascar. *Plant Systematics and Evolution* 285 (1-2):51-64.

Arnold, E. N. 2009. Relationships, evolution and biogeography of Semaphore geckos, Pristurus (Squamata, Sphaerodactylidae) based on morphology. *Zootaxa* (2060):1-21.

Audru, Jean-Christophe, Adnand Bitri, Jean-Francois Desprats, Pascal Dominique, Guillaume Eucher, Said Hachim, Olivier Jossot, Christian Mathon, Jean-Louis

Nedellec, Philippe Sabourault, Olivier Sedan, Philippe Stollsteiner, and Monique Terrier-Sedan. 2010. Major natural hazards in a tropical volcanic island: A review for Mayotte Island, Comoros archipelago, Indian Ocean. *Engineering Geology* 114 (3-4):364-381.

.udru, Jean-Christophe, Pol Guennoc, Isabelle Thinon, and Olivier Abellard. 2006. BATHYMAY: Underwater structure of Mayotte Island revealed by multibeam bathymetry. *Comptes Rendus Geoscience* 338 (16):1240-1249.

.ustin, J. J., E. N. Arnold, and C. G. Jones. 2004. Reconstructing an island radiation using ancient and recent DNA: the extinct and living day geckos (Phelsuma) of the Mascarene islands. *Molecular Phylogenetics and Evolution* 31 (1):109-122.

.ustin, J. J., E. N. Arnold, and C. G. Jones. 2009. Interrelationships and history of the slit-eared skinks (Gongylomorphus, Scincidae) of the Mascarene islands, based on mitochondrial DNA and nuclear gene sequences. *Zootaxa* (2153):55-68.

.ustin, JJ, EN Arnold, and CG Jones. 2004. Reconstructing an island radiation using ancient and recent DNA: the extinct and living day geckos (*Phelsuma*) of the Mascarene islands. *Molecular Phylogenetics and Evolution* 31:109-122.

.vise, J. C. 2004. What is the field of biogeography, and where is it going? *Taxon* 53 (4):893-898.

.vise, J. C., J. Arnold, R. M. Ball, E. Bermingham, T. Lamb, J. E. Neigel, C. A. Reeb, and N. C. Saunders. 1987. Intraspecific phylogeography - the mitochondrial-DNA bridge between population-genetics and systematics. *Annual Review of Ecology and Systematics* 18:489-522.

.vise, JC. 2000. *Phylogeography: the history and formation of species*: President and Fellows of Harvard College.

.artish, I. V., A. Antonelli, J. E. Richardson, and U. Swenson. 2011. Vicariance or long-distance dispersal: historical biogeography of the pantropical subfamily Chrysophylloideae (Sapotaceae). *Journal of Biogeography* 38 (1):177-190.

.ell, J.R., D.A. Bohan, E.M. Shaw, and G.S. Weyman. 2005. Ballooning dispersal using silk: world fauna, phylogenies, genetics and models. *Bull Entomol Res.* 95 (2):69-114.

.ellemain, E., E. Bermingham, and R. E. Ricklefs. 2008. The dynamic evolutionary history of the bananaquit (Coereba flaveola) in the Caribbean revealed by a multigene analysis. *Bmc Evolutionary Biology* 8:14.

.ellemain, E., and R. E. Ricklefs. 2008. Are islands the end of the colonization road? *Trends in Ecology & Evolution* 23 (8):461-468.

.iju, S. D., and F. Bossuyt. 2003. New frog family from India reveals an ancient biogeographical link with the Seychelles. *Nature* 425 (6959):711-714.

.lackledge, T. A., and R. G. Gillespie. 2004. Convergent evolution of behavior in an adaptive radiation of Hawaiian web-building spiders. *Proceedings of the National Academy of Sciences of the United States of America* 101 (46):16228-16233.

.one, T. S., S. R. Downie, J. M. Affolter, and K. Spalik. 2011. A Phylogenetic and Biogeographic Study of the Genus Lilaeopsis (Apiaceae tribe Oenantheae). *Systematic Botany* 36 (3):789-805.

.ossuyt, F., R. M. Brown, D. M. Hillis, D. C. Cannatella, and M. C. Milinkovitch. 2006. Phylogeny and biogeography of a cosmopolitan frog radiation: Late cretaceous diversification resulted in continent-scale endemism in the family ranidae. *Systematic Biology* 55 (4):579-594.

Bossuyt, F., and M. C. Milinkovitch. 2000. Convergent adaptive radiations in Madagascar and Asian ranid frogs reveal covariation between larval and adult traits. *Proceeding of the National Academy of Sciences of the United States of America* 97 (12):6585-6590.

Bourjea, J., S. Lapegue, L. Gagnevin, D. Broderick, J. A. Mortimer, S. Ciccione, D. Roos, C Taquet, and H. Grizel. 2007. Phylogeography of the green turtle, Chelonia mydas in the Southwest Indian Ocean. *Molecular Ecology* 16 (1):175-186.

Braby, M. F., J. W. H. Trueman, and R. Eastwood. 2005. When and where did troidin butterflies (Lepidoptera : Papilionidae) evolve? Phylogenetic and biogeographi evidence suggests an origin in remnant Gondwana in the Late Cretaceous *Invertebrate Systematics* 19 (2):113-143.

Briggs, J. C. 2003. The biogeographic and tectonic history of India. *Journal of Biogeography* 3 (3):381-388.

Bryan, S. E., I. U. Peate, D. W. Peate, S. Self, D. A. Jerram, M. R. Mawby, J. S. Marsh, and J. A Miller. 2010. The largest volcanic eruptions on Earth. *Earth-Science Reviews* 102 (3 4):207-229.

Bryson, R. W., U. O. Garcia-Vazquez, and B. R. Riddle. 2011. Phylogeography of Middl American gophersnakes: mixed responses to biogeographical barriers across th Mexican Transition Zone. *Journal of Biogeography* 38 (8):1570-1584.

Buerki, S., P. P. Lowry, S. Andriambololonera, P. B. Phillipson, L. Vary, and M. W Callmander. 2011. How to kill two genera with one tree: clarifying generi circumscriptions in an endemic Malagasy clade of Sapindaceae. *Botanical Journal o the Linnean Society* 165 (3):223-234.

Burridge, C. P. 2000. Biogeographic history of geminate cirrhitoids (Perciformes Cirrhitoidea) with east-west allopatric distributions across southern Australia based on molecular data. *Global Ecology and Biogeography* 9 (6):517-525.

Busais, SMS. 2011. Taxonomy and Molecular Phylogeny of *Hemidactylus* in the mainland o Yemen (Class: Reptilia, Order: Squamata, Family: Gekkonidae), Fakultät fü Lebenswissenschaften, Technischen Universität Carolo-Wilhelmina zu Braunschweig, Braunschweig.

Byrne, M., D. A. Steane, L. Joseph, D. K. Yeates, G. J. Jordan, D. Crayn, K. Aplin, D. J Cantrill, L. G. Cook, M. D. Crisp, J. S. Keogh, J. Melville, C. Moritz, N. Porch, J. M K. Sniderman, P. Sunnucks, and P. H. Weston. 2011. Decline of a biome: evolution contraction, fragmentation, extinction and invasion of the Australian mesic zon biota. *Journal of Biogeography* 38 (9):1635-1656.

Caccone, A., G. Amato, O. C. Gratry, J. Behler, and J. R. Powell. 1999. A molecula phylogeny of four endangered Madagascar tortoises based on MtDNA sequences *Molecular Phylogenetics and Evolution* 12 (1):1-9.

Caceres, S. 2003. *Étude préalable pour le classement en réserve naturelle des îles Éparses, mémoir de DESS Sciences et Gestion de l'Environnement tropical.* Saint-Denis: DIREN Réunion / Laboratoire d'écologie marine de l'Université de La Réunion.

Camacho-Garcia, Y. E., and T. M. Gosliner. 2008. Systematic revision of Jorunna Bergh, 187€ (Nudibranchia : Discodorididae) with a morphological phylogenetic analysis *Journal of Molluscan Studies* 74:143-181.

Carlquist, S. 1966. The biota of long-distance dispersal. I. Principles of dispersal and evolution. The Quarterly Review of Biology 41:247–270.

Carranza, S., and E. N. Arnold. 2006. Systematics, biogeography, and evolution of Hemidactylus geckos (Reptilia : Gekkonidae) elucidated using mitochondrial DNA sequences. *Molecular Phylogenetics and Evolution* 38 (2):531-545.

Catry, P., R. Mellanby, K. A. Suleiman, K. H. Salim, M. Hughes, M. McKean, N. Anderson, G. Constant, V. Heany, G. Martin, M. Armitage, and M. Wilson. 2000. Habitat selection by terrestrial birds on Pemba Island (Tanzania), with particular reference to six endemic taxa. *Biological Conservation* 95 (3):259-267.

Chakrabarty, P. 2004. Cichlid biogeography: comment and review. *Fish and Fisheries* 5 (2):97-119.

Chan, L. M., J. L. Brown, and A. D. Yoder. 2011. Integrating statistical genetic and geospatial methods brings new power to phylogeography. *Molecular Phylogenetics and Evolution* 59 (2):523-537.

Charruau, P., C. Fernandes, P. Orozco-ter Wengel, J. Peters, L. Hunter, H. Ziaie, A. Jourabchian, H. Jowkar, G. Schaller, S. Ostrowski, P. Vercammen, T. Grange, C. Schlotterer, A. Kotze, E. M. Geigl, C. Walzer, and P. A. Burger. 2011. Phylogeography, genetic structure and population divergence time of cheetahs in Africa and Asia: evidence for long-term geographic isolates. *Molecular Ecology* 20 (4):706-724.

Cheke, A. 2010. The timing of arrival of humans and their commensal animals on Western Indian Ocean oceanic islands. *Phelsuma* 18:38-69.

Cheke, A, and J Hume. 2008. *Lost land of the Dodo: an ecological history of Mauritius, Reunion and Rodrigues*. London: T & A D Poyser.

Clark, T. P. 1994. The species of *Walsura* and *Pseudoclausena* genus-novum (Meliaceae). *Blumea* 38 (2):247-302.

Clausnitzer, V. 2003. Teinobasis alluaudi Martin, 1896 from mainland Africa: Notes on ecology and biogeography (Zygoptera : Coenagrionidae). *Odonatologica* 32 (4):321-334.

Cooper, A., C. Lalueza-Fox, S. Anderson, A. Rambaut, J. Austin, and R. Ward. 2001. Complete mitochondrial genome sequences of two extinct moas clarify ratite evolution. *Nature* 409 (6821):704-707.

Corronca, J. A. 2003. New genus and species of Selenopidae (Arachnida, Araneae) from Madagascar and neighbouring islands. *African Zoology* 38 (2):387-392.

Cowie, R. H., and B. S. Holland. 2006. Dispersal is fundamental to biogeography and the evolution of biodiversity on oceanic islands. *Journal of Biogeography* 33 (2):193-198.

Cowie, R. H., and B. S. Holland. 2008. Molecular biogeography and diversification of the endemic terrestrial fauna of the Hawaiian Islands. *Philosophical Transactions of the Royal Society B-Biological Sciences* 363 (1508):3363-3376.

Coyle, F. A. 1995. A revision of the funnel web mygalomorph spider subfamily Ischnothelinae (Araneae, Dipluridae). *Bulletin of the American Museum of Natural History* (226):3-133.

Crisp, M. D., M. T. K. Arroyo, L. G. Cook, M. A. Gandolfo, G. J. Jordan, M. S. McGlone, P. H. Weston, M. Westoby, P. Wilf, and H. P. Linder. 2009. Phylogenetic biome conservatism on a global scale. *Nature* 458 (7239):754-U90.

Cuenca, A., C. B. Asmussen-Lange, and F. Borchsenius. 2008. A dated phylogeny of the palm tribe Chamaedoreeae supports Eocene dispersal between Africa, North and South America. *Molecular Phylogenetics and Evolution* 46 (2):760-775.

Cumberlidge, N. 2008. Insular species of Afrotropical freshwater crabs (Crustacea: Decapoda: Brachyura: Potamonautidae and Potamidae) with special reference to Madagascar and the Seychelles. *Contributions to Zoology* 77 (2):71-81.

Daniels, S. R. 2011. Reconstructing the colonisation and diversification history of the endemic freshwater crab (Seychellum alluaudi) in the granitic and volcanic Seychelles Archipelago. *Molecular Phylogenetics and Evolution* 61:534-542.

Daniels, S. R., N. Cumberlidge, M. Perez-Losada, S. A. E. Marijnissen, and K. A. Crandall. 2006. Evolution of Afrotropical freshwater crab lineages obscured by morphological convergence. *Molecular Phylogenetics and Evolution* 40 (1):227-235.

Darwin, CR. 1859. *On the origin of species by means of natural selection, or the preservation of favoured races in the struggle for life, 1st edition*. London: John Murray.

Darwin, CR, and AR Wallace. 1858. On the tendency of species to form varieties; and on the perpetuation of varieties and species by natural means of selection. *Proceedings of the Linnean Society of London. Zoology* 3:45-62.

de Queiroz, A. 2005. The resurrection of oceanic dispersal in historical biogeography. *Trends in Ecology & Evolution* 20 (2):68-73.

Deniel, C, G Kieffer, and J Lecointre. 1992. New 230Th-238U and 14C age determinations from Piton des Neiges volcano, Réunion - A revised chronology for the differentiated series. *Journal of Volcanolical and Geothermal Research* 51:253-267.

Dick, C. W., K. Abdul-Salim, and E. Bermingham. 2003. Molecular systematic analysis reveals cryptic tertiary diversification of a widespread tropical rain forest tree. *American Naturalist* 162 (6):691-703.

Dijkstra, K. D. B. 2007. Gone with the wind: westward dispersal across the Indian Ocean and island speciation in Hemicordulia dragonflies (Odonata : Corduliidae). *Zootaxa* (1438):27-48.

Dijkstra, K. D. B., V. Clausnitzer, and A. Martens. 2007. Tropical African Platycnemis damselflies (Odonata : Piatycnemididae) and the biogeographical significance of a new species from Pemba Island, Tanzania. *Systematics and Biodiversity* 5 (2):187-198.

Drummond, Alexei J., Simon Y. W. Ho, Matthew J. Phillips, and Andrew Rambaut. 2006. Relaxed phylogenetics and dating with confidence. *Plos Biology* 4 (5):699-710.

Dunbar-Co, S., A. M. Wieczorek, and C. W. Morden. 2008. Molecular phylogeny and adaptive radiation of the endemic Hawaiian Plantago species (Plantaginaceae). *American Journal of Botany* 95 (9):1177-1188.

Duncan, R. A., and m. Storey. 1992. The life cycle of Indian Ocean hotspots. *Geophysical Monograph* 70: 91-103.

Emerick, C. M., and R. A. Duncan. 1982. Age progressive volcanism in the Comores Archipelago, Western Indian Ocean and implications for Somali plate tectonics. *Earth and Planetary Science Letters* 60: 415-428.

Emerson, B. C. 2008. Speciation on islands: what are we learning? *Biological Journal of the Linnean Society* 95 (1):47-52.

Fijarczyk, A., K. Nadachowska, S. Hofman, S. N. Litvinchuk, W. Babik, M. Stuglik, G. Gollmann, L. Choleva, D. Cogalniceanu, T. Vukov, G. Dzukic, and J. M. Szymura. 2011. Nuclear and mitochondrial phylogeography of the European fire-bellied toads Bombina bombina and Bombina variegata supports their independent histories. *Molecular Ecology* 20 (16):3381-3398.

isher, B. L. 2007. Biogeography and ecology of the ant fauna of Madagascar (Hymenoptera: Formicidae). *Journal of Natural History* 31 (2): 269-302.

uchs, J., J. M. Pons, S. M. Goodman, V. Bretagnolle, M. Melo, R. C. K. Bowie, D. Currie, R. Safford, M. Z. Virani, S. Thomsett, A. Hija, C. Cruaud, and E. Pasquet. 2008. Tracing the colonization history of the Indian Ocean scops-owls (Strigiformes : Otus) with further insight into the spatio-temporal origin of the Malagasy avifauna. *Bmc Evolutionary Biology* 8.

uller, S, M Schwarz, and S Tierney. 2005. Phylogenetics of the allodapine bee genus Braunsapis: historical biogeography and long-range dispersal over water. *Journal of Biogeography* 32:2135-2144.

Gage, E., P. Wilkin, M. W. Chase, and J. Hawkins. 2011. Phylogenetic systematics of Sternbergia (Amaryllidaceae) based on plastid and ITS sequence data. *Botanical Journal of the Linnean Society* 166 (2):149-162.

Gamble, T., A. M. Bauer, E. Greenbaum, and T. R. Jackman. 2008. Evidence for Gondwanan vicariance in an ancient clade of gecko lizards. *Journal of Biogeography* 35 (1):88-104.

Garb, J. E., and R. G. Gillespie. 2009. Diversity despite dispersal: colonization history and phylogeography of Hawaiian crab spiders inferred from multilocus genetic data. *Molecular Ecology* 18 (8):1746-1764.

Gillespie, R. G. 2004. Community assembly through adaptive radiation in Hawaiian spiders. *Science* 303:356-359.

Gillespie, R. G. 2005. Geographical context of speciation in a radiation of Hawaiian Tetragnatha spiders (Araneae, Tetragnathidae). *Journal of Arachnology* 33 (2):313-322.

Gillespie, R. G., B. G. Baldwin, J. M. Waters, C. I. Fraser, R. Nikula, and G. K. Roderick. 2012. Long-distance dispersal: a framework for hypothesis testing. Trends in Ecology and Evolution 27:52-61.

Gillespie, R. G., E. M. Claridge, and S. L. Goodacre. 2008. Biogeography of the fauna of French Polynesia: diversification within and between a series of hot spot archipelagos. *Philosophical Transactions of the Royal Society B-Biological Sciences* 363 (1508):3335-3346.

Gillespie, R. G., and G. K. Roderick. 2002. Arthropods on islands: Colonization, speciation, and conservation. *Annual Review of Entomology* 47:595-632.

Gillot, P.Y., J. C. Lefèvre, and P. E. Nativel. 1994. Model for the structural evolution of the volcanoes of Réunion Island. *Earth and Planetary Science Letters* 122: 291-302.

Goldberg, E. E., L. T. Lancaster, and R. H. Ree. 2011. Phylogenetic Inference of Reciprocal Effects between Geographic Range Evolution and Diversification. *Systematic Biology* 60 (4):451-465.

Goodman, S. M., W. Buccas, T. Naidoo, F. Ratrimomanarivo, P. J. Taylor, and J. Lamb. 2010. Patterns of morphological and genetic variation in western Indian Ocean members of the *Chaerephon* 'pumilus' complex (Chiroptera: Molossidae), with the description of a new species from Madagascar. *Zootaxa* (2551):1-36.

Goodman, S. M., L. M. Chan, M. D. Nowak, and A. D. Yoder. 2010. Phylogeny and biogeography of western Indian Ocean *Rousettus* (Chiroptera: Pteropodidae). *Journal of Mammalogy* 91 (3):593-606.

Gower, David J., Diego San Mauro, Varad Giri, Gopalakrishna Bhatta, Venu Govindappa, Ramachandran Kotharambath, Oommen V. Oommen, Farrah A. Fatih, Jacqueline A. Mackenzie-Dodds, Ronald A. Nussbaum, S. D. Biju, Yogesh S. Shouche, and

Mark Wilkinson. 2011. Molecular systematics of caeciliid caecilians (Amphibia Gymnophiona) of the Western Ghats, India. *Molecular Phylogenetics and Evolution* 59 (3):698-707.

Grande, L. 1985. The use of paleontology in systematics and biogeography, and a time control refinement for historical biogeography. *Paleobiology* 11:234-243.

Griswold, C. E. 1997. The spider family Cyatholipidae in Madagascar (Araneae Araneoidea). *Journal of Arachnology* 25 (1):53-83.

Griswold, C, and J. Ledford. 2001. A monograph of the migid trap-door spiders of Madagascar, with a phylogeny of world genera (Araneae, Mygalomorphae, Migidae). Occasional Papers of the California Academy of Sciences 151:1-120.

Groombridge, J. J., C. G. Jones, M. K. Bayes, A. J. van Zyl, J. Carrillo, R. A. Nichols, and M. W. Bruford. 2002. A molecular phylogeny of African kestrels with reference to divergence across the Indian Ocean. *Molecular Phylogenetics and Evolution* 25 (2):267-277.

Hare, M. P., and J. C. Avise. 1998. Population structure in the American oyster as inferred by nuclear gene genealogies. *Molecular Biology and Evolution* 15 (2):119-128.

Harris, D. J., and S. Rocha. 2009. Comoros. In *Encyclopedia of Islands*, edited by R. C. Gillespie and D. A. Clague. Berkeley: University of California Press, 612-619.

Harmon, L. J., J. Melville, A. Larson, and J. B. Losos. 2008. The Role of Geography and Ecological Opportunity in the Diversification of Day Geckos (Phelsuma). *Systematic Biology* 57 (4):562-573.

Heads, M. 2009. Globally basal centres of endemism: the Tasman-Coral Sea region (southwest Pacific), Latin America and Madagascar/South Africa. *Biological Journal of the Linnean Society* 96 (1):222-245.

Heads, M. 2011. Old Taxa on Young Islands: A Critique of the Use of Island Age to Date Island-Endemic Clades and Calibrate Phylogenies. *Systematic Biology* 60 (2):204-U142.

Heaney, L. R. 2007. Is a new paradigm emerging for oceanic island biogeography? *Journal of Biogeography* 34 (5):753-757.

Hedges, S. B., and M. P. Heinicke. 2007. Molecular phylogeny and biogeography of west Indian frogs of the genus *Leptodactylus* (Anura, Leptodactylidae). *Molecular Phylogenetics and Evolution* 44 (1):308-314.

Herrmann, M., S. Kienle, J. Rochat, W. E. Mayer, and R. J. Sommer. 2010. Haplotype diversity of the nematode *Pristionchus pacificus* on Reunion in the Indian Ocean suggests multiple independent invasions. *Biological Journal of the Linnean Society* 100 (1):170-179.

Hickerson, M. J., B. C. Carstens, J. Cavender-Bares, K. A. Crandall, C. H. Graham, J. B. Johnson, L. Rissler, P. F. Victoriano, and A. D. Yoder. 2010. Phylogeography's past, present, and future: 10 years after Avise, 2000. *Molecular Phylogenetics and Evolution* 54 (1):291-301.

Holland, B. S., and R. H. Cowie. 2006. Dispersal and vicariance in Hawaii: submarine slumping does not create deep inter-island channels. *Journal of Biogeography* 33 (12):2155-2156.

Hommersand, M. H., S. Fredericq, and D. W. Freshwater. 1994. Phylogenetic systematics and biogeography of the Gigartinaceae (Gigartinales, Rhodophyta) Based on Sequence-Analysis Of Rbcl. *Botanica Marina* 37 (3):193-203.

luber, B. A. 2003. Cladistic analysis of Malagasy pholcid spiders reveals generic level endemism: Revision of *Zatavua* n. gen. and *Paramicromerys* Millot (Pholcidae, Araneae). *Zoological Journal of the Linnean Society* 137 (2):261-318.

luber, B. A., and H. K. El-Hennawy. 2007. On Old World ninetine spiders (Araneae : Pholcidae), with a new genus and species and the first record for Madagascar. *Zootaxa* (1635):45-53.

lume, J. P. 2007. Reappraisal of the parrots (Aves : Psittacidae) from the Mascarene Islands, with comments on their ecology, morphology, and affinities. *Zootaxa* (1513):3-76.

lunn, CA, and P. Upchurch. 2001. The importance of time/space in diagnosing the causality of phylogenetic events: towards a "chronobiogeographical" paradigm? *Systematic Biology* 50:391–407.

UCN. *The IUCN red list of threatened species.* http://www.iucnredlist.org/ 2011.

aquemet, S., M. Le Corre, and G. D. Quartly. 2007. Ocean control of the breeding regime of the sooty tern in the southwest Indian Ocean. *Deep-Sea Research Part I-Oceanographic Research Papers* 54 (1):130-142.

onsson, KA, and J Fjeldsa. 2006. Determining biogeographical patterns of dispersal and diversification in oscine passerine birds in Australia, Southeast Asia and Africa. *Journal of Biogeography* 33:1155-1165.

Kensley, B., and M. Schotte. 2000. New species and records of anthuridean isopod crustaceans from the Indian Ocean. *Journal of Natural History* 34 (11):2057-2121.

Kerdelhue, C., I. Le Clainche, and J. Y. Rasplus. 1999. Molecular phylogeny of the Ceratosolen species pollinating Ficus of the subgenus Sycomorus sensu stricto: Biogeographical history and origins of the species-specificity breakdown cases. *Molecular Phylogenetics and Evolution* 11 (3):401-414.

Kingdon, J. 1989. *Island Africa.* Princeton: Princeton University Press.

Kisel, Y., and T. G. Barraclough. 2010. Speciation has a spatial scale that depends on levels of gene flow. *American Naturalist* 175 (3):316-334.

Kita, Y., and M. Kato. 2004. Phylogenetic relationships between disjunctly occurring groups of Tristicha trifaria (Podostemaceae). *Journal of Biogeography* 31 (10):1605-1612.

Klaus, S., C. D. Schubart, and D. Brandis. 2006. Phylogeny, biogeography and a new taxonomy for the Gecarcinucoidea Rathbun, 1904 (Decapoda : Brachyura). *Organisms Diversity & Evolution* 6 (3):199-217.

Knoflach, B., and A. Van Harten. 2006. The one-palped spider genera Tidarren and Echinotheridion in the Old World (Araneae, Theridiidae), with comparative remarks on Tidarren from America. *Journal of Natural History* 40 (25-26):1483-1616.

Knowles, L. L., and W. P. Maddison. 2002. Statistical phylogeography. *Molecular Ecology* 11 (12):2623-2635.

Kock, D., and W. T. Stanley. 2009. Mammals of Mafia Island, Tanzania. *Mammalia* 73 (4):339-352.

Kohler, F., and M. Glaubrecht. 2010. Uncovering an overlooked radiation: molecular phylogeny and biogeography of Madagascar's endemic river snails (Caenogastropoda: Pachychilidae: *Madagasikara* gen. nov.). *Biological Journal of the Linnean Society* 99 (4):867-894.

Krause, DW. 2003. Late Cretaceous vertebrates of Madagascar: a window into Gondwanan biogeography at the end of the age of dinosaurs. In *The Natural History of*

Madagascar, edited by S. Goodman and J. Benstead. ChicagoUniversity of Chicago Press.

Kulju, K. K. M., S. E. C. Sierra, S. G. A. Draisma, R. Samuel, and P. C. van Welzen. 2007. Molecular phylogeny of *Macaranga, Mallotus*, and related genera (Euphorbiaceae s.s.): insights from plastid and nuclear DNA sequence data. *American Journal of Botany* 94 (10):1726-1743.

Kuntner, M. 2006. Phylogenetic systematics of the Gondwanan nephilid spider lineage Clitaetrinae (Araneae, Nephilidae). *Zoologica Scripta* 35 (1):19-62.

Kuntner, M. 2007. A monograph of *Nephilengys*, the pantropical 'hermit spiders' (Araneae Nephilidae, Nephilinae). *Systematic Entomology* 32 (1):95-135.

Kuntner, M., and I. Agnarsson. 2009. Phylogeny accurately predicts behaviour in Indian Ocean *Clitaetra* spiders (Araneae : Nephilidae). *Invertebrate Systematics* 23 (3):193-204.

Kuntner, M., and I. Agnarsson. 2011a. Biogeography and diversification of hermit spiders on Indian Ocean islands (Nephilidae: *Nephilengys*). *Molecular Phylogenetics and Evolution* 59 (2):477-488.

Kuntner, M., and I. Agnarsson. 2011b. Phylogeography of a successful aerial disperser: the golden orb spider *Nephila* on Indian Ocean islands. *BMC Evolutionary Biology* 11.

Kuntner, M., J. A. Coddington, and G. Hormiga. 2008. Phylogeny of extant nephilid orb-weaving spiders (Araneae, Nephilidae): testing morphological and ethologica homologies. *Cladistics* 24 (2):147-217.

Kuntner, M., C. R. Haddad, G. Aljančič, and A. Blejec. 2008. Ecology and web allometry of *Clitaetra irenae*, an arboricolous African orb-weaving spider (Araneae, Araneoidea Nephilidae). *Journal of Arachnology* 36 (3):583-594.

Kuntner, M., S. Kralj-Fišer, and M. Gregorič. 2010. Ladder webs in orb-web spiders ontogenetic and evolutionary patterns in Nephilidae. *Biological Journal of the Linnean Society* 99 (4):849-866.

Le Pechon, T., N. Cao, J. Y. Dubuisson, and L. D. B. Gigord. 2009. Systematics of Dombeyoideae (Malvaceae) in the Mascarene archipelago (Indian Ocean) inferred from morphology. *Taxon* 58 (2):519-531.

Le Pechon, T., J. Y. Dubuisson, T. Haevermans, C. Cruaud, A. Couloux, and L. D. B. Gigord 2010. Multiple colonizations from Madagascar and converged acquisition of dioecy in the Mascarene Dombeyoideae (Malvaceae) as inferred from chloroplast and nuclear DNA sequence analyses. *Annals of Botany* 106 (2):343-357.

Lehtonen, Samuli, Hanna Tuomisto, Germinal Rouhan, and Maarten J. M. Christenhusz. 2010. Phylogenetics and classification of the pantropical fern family Lindsaeaceae. *Botanical Journal of the Linnean Society* 163 (3):305-359.

Leigh, EG, A Hladik, CM Hladik, and A Jolly. 2007. The biogeography of large islands, or how does the size of the ecological theater affect the evolutionary play? *Revue D Ecologie-La Terre Et La Vie* 62 (2-3):105-168.

Logunov, D. V. 2009. Further notes on the Harmochireae of Africa (Araneae, Salticidae, Pelleninae). *Zookeys* (16):265-290.

Losos, J. B. 1988. Ecomorphological Evolution in West-Indian *Anolis* Lizards. *American Zoologist* 28 (4):A15-A15.

Losos, J. B., and K. DeQueiroz. 1997. Evolutionary consequences of ecological release in Caribbean *Anolis* lizards. *Biological Journal of the Linnean Society* 61 (4):459-483.

Lundberg, JG. 1993. African-South American freshwater fish clades and continental drift: problems with a paradigm. In *Biological Relationships between Africa and South America*, edited by P. Goldblatt. New Haven: Yale University Press.

Lyle, R., and C. R. Haddad. 2010. A revision of the tracheline sac spider genus Cetonana Strand, 1929 in the Afrotropical Region, with descriptions of two new genera (Araneae: Corinnidae). *African Invertebrates* 51 (2):321-384.

Macey, J. Robert, Jennifer V. Kuehl, Allan Larson, Michael D. Robinson, Ismail H. Ugurtas, Natalia B. Ananjeva, Hafizur Rahman, Hamid Iqbal Javed, Ridwan Mohamed Osmani, Ali Doumma, and Theodore J. Papenfuss. 2008. Socotra Island the forgotten fragment of Gondwana: Unmasking chameleon lizard history with complete mitochondrial genomic data. *Molecular Phylogenetics and Evolution* 49 (3):1015-1018.

Maddison, W. P. 1997. Gene trees in species trees. *Systematic Biology* 46 (3):523-536.

Maddison, WP. 1995. Phylogenetic histories within and among species. In *Experimental and molecular approaches to plant biosystematics. Monographs in Systematics*, edited by P. Hoch and A. Stevenson. St. Louis: Missouri Botanical Garden.

Manns, U., and A. A. Anderberg. 2011. Biogeography of 'tropical Anagallis' (Myrsinaceae) inferred from nuclear and plastid DNA sequence data. *Journal of Biogeography* 38 (5):950-961.

Marks, B. D., and D. E. Willard. 2005. Phylogenetic relationships of the Madagascar Pygmy Kingfisher (Ispidina madagascariensis). *Auk* 122 (4):1271-1280.

Martin, R. D. 2000. Origins, diversity and relationships of lemurs. *International Journal of Primatology* 21 (6):1021-1049.

Masters, J. C., M. J. de Wit, and R. J. Asher. 2006. Reconciling the origins of Africa, India and Madagascar with vertebrate dispersal scenarios. *Folia Primatologica* 77 (6):399-418.

Masters, J. C., B. G. Lovegrove, and M. J. de Wit. 2007. Eyes wide shut: can hypometabolism really explain the primate colonization of Madagascar? *Journal of Biogeography* 34 (1):21-37.

Maurin, O., A. P. Davis, M. Chester, E. F. Mvungi, Y. Jaufeerally-Fakim, and M. F. Fay. 2007. Towards a phylogeny for Coffea (Rubiaceae): Identifying well-supported lineages based on nuclear and plastid DNA sequences. *Annals of Botany* 100 (7):1565-1583.

Mausfeld, P., M. Vences, A. Schmitz, and M. Veith. 2000. First data on the molecular phylogeography of scincid lizards of the genus Mabuya. *Molecular Phylogenetics and Evolution* 17 (1):11-14.

Miller, AG, and M Miranda. 2004. *Flora of the Soqotra Archipelago*. Edinburgh: Royal Botanic Garden.

Morgan, WJ. 1981. Hotspot tracks and the opening of the Atlantic and Indian oceans. In *The Sea*, edited by C. Emiliani. New York: Wiley.

Msuya, C. A., K. M. Howell, and A. Channing. 2006. A new species of Running Frog, (Kassina, Anura : Hyperoliidae) from Unguja Island, Zanzibar, Tanzania. *African Journal of Herpetology* 55 (2):113-122.

Nagy, Z. T., U. Joger, M. Wink, F. Glaw, and M. Vences. 2003. Multiple colonization of Madagascar and Socotra by colubrid snakes: evidence from nuclear and mitochondrial gene phylogenies. *Proceedings of the Royal Society of London Series B-Biological Sciences* 270 (1533):2613-2621.

Nathan, R. 2006. Long-distance dispersal of plants. *Science* 313:786–788.

Nazari, V., T. B. Larsen, D. C. Lees, O. Brattstrom, T. Bouyer, G. Van de Poel, and P. D. N. Hebert. 2011. Phylogenetic systematics of Colotis and associated genera (Lepidoptera: Pieridae): evolutionary and taxonomic implications. *Journal of Zoological Systematics and Evolutionary Research* 49 (3):204-215.

Negi, J. G., P. K. Agrawal, O. P. Pandey, and A. P. Singh. 1993. A possible K-T boundary bolide impact site offshore near Bombay and triggering of rapid Deccan volcanism. *Physics of the Earth and Planetary Interiors* 76 (3-4):189-197.

Nelson, G. 1979. From Candolle to Croizat: comments on the history of biogeography. *Journal of Historical Biology* 11:269-305.

Nelson, J. S., and N. I. Platnick. 1981. *Systematics and biogeography. Cladistics and vicariance.* New York: Columbia University Press.

Nougier, J., J. M. Cantagrel, and J. P. Karche. 1986. The Comores archipelago in the western Indian Ocean: volcanology, geochronology and geodynamic setting. Journal of African Earth Sciences, Vol. 5, No. 2, pp. 135-145.

Noonan, B. P., and P. T. Chippindale. 2006. Vicariant origin of Malagasy reptiles supports late cretaceous antarctic land bridge. *American Naturalist* 168 (6):730-741.

Nowak, K., and P. C. Lee. 2011. Demographic Structure of Zanzibar Red Colobus Populations in Unprotected Coral Rag and Mangrove Forests. *International Journal of Primatology* 32 (1):24-45.

O'Brien, J., C. Mariani, L. Olson, A. L. Russell, L. Say, A. D. Yoder, and T. J. Hayden. 2009. Multiple colonisations of the western Indian Ocean by Pteropus fruit bats (Megachiroptera: Pteropodidae): The furthest islands were colonised first. *Molecular Phylogenetics and Evolution* 51 (2):294-303.

Olesen, J. 1999. A new species of Nebalia (Crustacea, Leptostraca) from Unguja Island (Zanzibar), Tanzania, East Africa, with a phylogenetic analysis of leptostracan genera. *Journal of Natural History* 33 (12):1789-1809.

Orsini, L., H. Koivulehto, and I. Hanski. 2007. Molecular evolution and radiation of dung beetles in Madagascar. *Cladistics* 23 (2):145-168.

Parent, C. E., A. Caccone, and K. Petren. 2008. Colonization and diversification of Galapagos terrestrial fauna: a phylogenetic and biogeographical synthesis. *Philosophical Transactions of the Royal Society B-Biological Sciences* 363:3347-3361.

Pearson, RG, and CJ Raxworthy. 2009. The evolution of local endemism in Madagascar: Watershed versus climatic gradient hypotheses evaluated by null biogeographic models. *Evolution* 63 (4):959-967.

Peters, J. L., Y. Zhuravlev, I. Fefelov, A. Logie, and K. E. Omland. 2007. Nuclear loci and coalescent methods support ancient hybridization as cause of mitochondrial paraphyly between gadwall and falcated duck (*Anas* spp.). *Evolution* 61 (8):1992-2006.

Pickford, M., A. Bhandari, S. Bajpai, B. N. Tiwari, and D. M. Mohabey. 2008. Miocene terrestrial mammals from Circum-Indian Ocean: Implications for geochronology, biogeography, eustacy and Himalayan orogenesis. *Himalayan Geology* 29 (3):71-72.

Platnick, N. I. 1995. New species and records of the ground spider family Gallieniellidae (Araneae, Gnaphosoidea) from Madagascar. *Journal of Arachnology* 23 (1):9-12.

Plummer, PS, and ER Belle. 1995. Mesozoic tectonostratigraphic evolution of the Seychelles microcontinent. *Sedimentary Geology* 96:73-91.

Pokorny, L., G. Olivan, and A. J. Shaw. 2011. Phylogeographic patterns in two southern hemisphere species of *Calyptrochaeta* (Daltoniaceae, Bryophyta). *Systematic Botany* 36 (3):542-553.

Poux, C., O. Madsen, E. Marquard, D. R. Vieites, W. W. de Jong, and M. Vences. 2005. Asynchronous Colonization of Madagascar by the four endemic clades of Primates, Tenrecs, Carnivores, and Rodents as inferred from nuclear genes. *Systematic Biology* 54 (5):719-730.

Rabinovitz, PD, MF Coffin, and D Falvey. 1983. The separation of Madagascar and Africa. *Science* 220:67-69.

Rage, J. C. 2003. Relationships of the Malagasy fauna during the Late Cretaceous: Northern or Southern routes? *Acta Palaeontologica Polonica* 48 (4):661-662.

Raxworthy, C. J., M. R. J. Forstner, and R. A. Nussbaum. 2002. Chameleon radiation by oceanic dispersal. *Nature* 415 (6873):784-787.

Raxworthy, C. J., C. M. Ingram, N. Rabibisoa, and R. G. Pearson. 2007. Applications of ecological niche modeling for species delimitation: A review and empirical evaluation using day geckos (Phelsuma) from Madagascar. *Systematic Biology* 56 (6):907-923.

Rehan, S. M., T. W. Chapman, A. I. Craigie, M. H. Richards, S. J. B. Cooper, and M. P. Schwarz. 2010. Molecular phylogeny of the small carpenter bees (Hymenoptera: Apidae: Ceratinini) indicates early and rapid global dispersal. *Molecular Phylogenetics and Evolution* 55 (3):1042-1054.

Renner, S. 2004. Plant dispersal across the tropical Atlantic by wind and sea currents. *International Journal of Plant Sciences* 165 (4):S23-S33.

Renner, S. S. 2004. Multiple Miocene Melastomataceae dispersal between Madagascar, Africa and India. *Philosophical Transactions of the Royal Society of London Series B-Biological Sciences* 359 (1450):1485-1494.

Renner, S. S., J. S. Strijk, D. Strasberg, and C. Thebaud. 2010. Biogeography of the Monimiaceae (Laurales): a role for East Gondwana and long-distance dispersal, but not West Gondwana. *Journal of Biogeography* 37 (7):1227-1238.

Ricklefs, R., and E. Bermingham. 2008. The West Indies as a laboratory of biogeography and evolution. *Philosophical Transactions of the Royal Society B-Biological Sciences* 363 (1502):2393-2413.

Ricklefs, R. E., and D. G. Jenkins. 2011. Biogeography and ecology: towards the integration of two disciplines. *Philosophical Transactions of the Royal Society B-Biological Sciences* 366 (1576):2438-2448.

Robinson, J. E., D. J. Bell, F. M. Saleh, A. A. Suleiman, and I. Barr. 2010. Recovery of the Vulnerable Pemba flying fox *Pteropus voeltzkowi*: population and conservation status. *Oryx* 44 (3):416-423.

Rocha, S, M Vences, F Glaw, D. Posada, and DJ Harris. 2009. Multigene phylogeny of Malagasy day geckos of the genus *Phelsuma*. *Molecular Phylogenetics and Evolution* 52:530-537.

Rocha, S., M. A. Carretero, and D. J. Harris. 2005. Mitochondrial DNA sequence data suggests two independent colonizations of the Comoros archipelago-by Chameleons of the genus *Furcifer*. *Belgian Journal of Zoology* 135 (1):39-42.

Rocha, S., M. A. Carretero, M. Vences, F. Glaw, and D. J. Harris. 2006. Deciphering patterns of transoceanic dispersal: the evolutionary origin and biogeography of coastal lizards (*Cryptoblepharus*) in the Western Indian Ocean region. *Journal of Biogeography* 33 (1):13-22.

Rocha, S., D. J. Harris, and D. Posada. 2011. Cryptic diversity within the endemic prehensile-tailed gecko *Urocotyledon inexpectata* across the Seychelles Islands: patterns of phylogeographical structure and isolation at the multilocus level. *Biological Journal of the Linnean Society* 104 (1):177-191.

Rocha, S., D. Posada, M. A. Carretero, and D. J. Harris. 2007. Phylogenetic affinities of Comoroan and East African day geckos (genus *Phelsuma*): Multiple natural colonisations, introductions and island radiations. *Molecular Phylogenetics and Evolution* 43 (2):685-692.

Rodder, D., O. Hawlitschek, and F. Glaw. 2010. Environmental niche plasticity of the endemic gecko *Phelsuma parkeri* Loveridge 1941 from Pemba Island, Tanzania: a case study of extinction risk on flat islands by climate change. *Tropical Zoology* 23 (1):35-49.

Rösler, H, and W Wranik. 2004. A key and annotated checklist to the reptiles of the Socotra Archipelago. *Fauna of Arabia* 20:505-534.

Rothe, N., A. J. Gooday, T. Cedhagen, and J. A. Hughes. 2011. Biodiversity and distribution of the genus *Gromia* (Protista, Rhizaria) in the deep Weddell Sea (Southern Ocean). *Polar Biology* 34 (1):69-81.

Rowson, B. 2007. Land molluscs of Zanzibar Island (Unguja), Tanzania, including a new species of Gulella (Pulmonata : Streptaxidae). *Journal of Conchology* 39:425-466.

Rowson, B., P. Tattersfield, and W. O. C. Symondson. 2011. Phylogeny and biogeography of tropical carnivorous land-snails (Pulmonata: Streptaxoidea) with particular reference to East Africa and the Indian Ocean. *Zoologica Scripta* 40 (1):85-98.

Rowson, B., B. H. Warren, and C. F. Ngereza. 2010. Terrestrial molluscs of Pemba Island, Zanzibar, Tanzania, and its status as an "oceanic" island. *Zookeys* (70):1-39.

Russell, AL, SM Goodman, and MP Cox. 2008. Coalescent analyses support multiple mainland-to-island dispersals in the evolution of Malagasy *Triaenops* bats (Chiroptera: Hipposideridae). *Journal of Biogeography* 35:995-1003.

Sanderson, M. J. 2002. Estimating absolute rates of molecular evolution and divergence times: A penalized likelihood approach. *Molecular Biology and Evolution* 19 (1):101-109.

Schaefer, H., C. Heibl, and S. S. Renner. 2009. Gourds afloat: a dated phylogeny reveals an Asian origin of the gourd family (Cucurbitaceae) and numerous oversea dispersal events. *Proceedings of the Royal Society B-Biological Sciences* 276 (1658):843-851.

Schatz, GE. 1996. Malagasy/Indo-australo-malesian phytogeographic connections. In *Biogéographie de Madagascar*, edited by W. Lourenço. Paris: ORSTOM.

Schluter, D. 2000. *The ecology of adaptive radiation.* New York: Oxford University Press.

Schluter, Dolph, and Laura M. Nagel. 1995. Parallel speciation by natural selection. *American Naturalist* 146 (2):292-301.

Schlüter, Thomas. 2006. The Comoros (Mayotte is still under French administration). Geological Atlas of Africa: Springer Berlin Heidelberg.

Schulte, P., L. Alegret, I. Arenillas, J. A. Arz, P. J. Barton, P. R. Bown, T. J. Bralower, G. L. Christeson, P. Claeys, C. S. Cockell, G. S. Collins, A. Deutsch, T. J. Goldin, K. Goto, J. M. Grajales-Nishimura, R. A. F. Grieve, S. P. S. Gulick, K. R. Johnson, W. Kiessling, C. Koeberl, D. A. Kring, K. G. MacLeod, T. Matsui, J. Melosh, A. Montanari, J. V. Morgan, C. R. Neal, D. J. Nichols, R. D. Norris, E. Pierazzo, G.

Ravizza, M. Rebolledo-Vieyra, W. U. Reimold, E. Robin, T. Salge, R. P. Speijer, A. R. Sweet, J. Urrutia-Fucugauchi, V. Vajda, M. T. Whalen, and P. S. Willumsen. 2010. The chicxulub asteroid impact and mass extinction at the Cretaceous-Paleogene boundary. *Science* 327 (5970):1214-1218.

chwarz, M. P., S. Fuller, S. M. Tierney, and S. J. B. Cooper. 2006. Molecular phylogenetics of the exoneurine allodapine bees reveal an ancient and puzzling dispersal from Africa to Australia. *Systematic Biology* 55 (1):31-45.

hapiro, B., D. Sibthorpe, A. Rambaut, J. Austin, G. M. Wragg, O. R. P. Bininda-Emonds, P. L. M. Lee, and A. Cooper. 2002. Flight of the dodo. *Science* 295 (5560):1683-1683.

heldon, F. H., D. J. Lohman, H. C. Lim, F. Zou, S. M. Goodman, D. M. Prawiradilaga, K. Winker, T. M. Braile, and R. G. Moyle. 2009. Phylogeography of the magpie-robin species complex (Aves: Turdidae: Copsychus) reveals a Philippine species, an interesting isolating barrier and unusual dispersal patterns in the Indian Ocean and Southeast Asia. *Journal of Biogeography* 36 (6):1070-1083.

hih, H. T., D. C. J. Yeo, and P. K. L. Ng. 2009. The collision of the Indian plate with Asia: molecular evidence for its impact on the phylogeny of freshwater crabs (Brachyura: Potamidae). *Journal of Biogeography* 36 (4):703-719.

mykal, P., G. Kenicer, A. J. Flavell, J. Corander, O. Kosterin, R. J. Redden, R. Ford, C. J. Coyne, N. Maxted, M. J. Ambrose, and N. T. H. Ellis. 2011. Phylogeny, phylogeography and genetic diversity of the *Pisum* genus. *Plant Genetic Resources-Characterization and Utilization* 9 (1):4-18.

ole, C. L., and C. H. Scholtz. 2010. Did dung beetles arise in Africa? A phylogenetic hypothesis based on five gene regions. *Molecular Phylogenetics and Evolution* 56 (2):631-641.

parks, J. S. 2004. Molecular phylogeny and biogeography of the Malagasy and South Asian cichlids (Teleostei : Perciformes : Cichlidae). *Molecular Phylogenetics and Evolution* 30 (3):599-614.

parks, J. S., and R. C. Schelly. 2011. A new species of *Paretroplus* (Teleostei: Cichlidae: Etroplinae) from northeastern Madagascar, with a phylogeny and revised diagnosis for the *P. damii* clade. *Zootaxa* (2768):55-68.

tanley, W. T. 2008. A new species of Mops (Molossidae) from Pemba Island, Tanzania. *Acta Chiropterologica* 10 (2):183-192.

teenis, CGGJ van. 1962. The land-bridge theory in botany. *Blumea* 11:235-372.

toddart, DR, ed. 1970. *Coral Islands of the western Indian Ocean.* Vol. 136, *Atoll Research Bulletin.* Washington DC: The Smithsonian Institution.

torey, M., J. J. Mahoney, A. D. Saunders, R. A. Duncan, S. P. Kelley, and M. F. Coffin. 1995. Timing of hot spot-related volcanism and the breakup of Madagascar and India. *Science* 267 (5199):852-855.

tothard, J. R., N. J. Loxton, and D. Rollinson. 2002. Freshwater snails on Mafia Island, Tanzania with special emphasis upon the genus *Bulinus* (Gastropoda : Planorbidae). *Journal of Zoology* 257:353-364.

tuart, SN, and Adams RJ. 1990. *Biodiversity in sub-Saharan Africa and its islands - conservation, management and sustainable use.* Vol. 6, *Occacional Papers of the IUCN Species Survival Commission.*

Sturge, R. J., F. Jacobsen, B. B. Rosensteel, R. J. Neale, and K. E. Omland. 2009. Colonization of South America from Caribbean islands confirmed by molecular phylogeny with increased taxon sampling. *Condor* 111 (3):575-579.

Tattersall, I. 2006. Historical biogeography of the strepsirhine primates of Madagascar. *Folia Primatologica* 77 (6):477-487.

Thébaud, C., B. H. Warren, D. Strasberg, and A. Cheke. 2009. Mascarene Islands, Biology. In *Encyclopedia of Islands*, edited by R. C. Gillespie and D. A. Clague. Berkeley: University of California Press, 612-619.

Thiv, M., and U. Meve. 2007. A phylogenetic study of *Echidnopsis* Hook. f. (Apocynaceae-Asclepiadoideae) - taxonomic implications and the colonization of the Socotran archipelago. *Plant Systematics and Evolution* 265 (1-2):71-86.

Thiv, M., M. Thulin, N. Kilian, and H. P. Linder. 2006. Eritreo-Arabian affinities of the Socotran flora as revealed from the molecular phylogeny of *Aerva* (Amaranthaceae). *Systematic Botany* 31 (3):560-570.

Thompson, A. 2000. *Origins of Arabia*. London: Stacey International.

Toon, A., M. Perez-Losada, C. E. Schweitzer, R. M. Feldmann, M. Carlson, and K. A. Crandall. 2010. Gondwanan radiation of the Southern Hemisphere crayfishes (Decapoda: Parastacidae): evidence from fossils and molecules. *Journal of Biogeography* 37 (12):2275-2290.

Townsend, T. M., K. A. Tolley, F. Glaw, W. Bohme, and M. Vences. 2011. Eastward from Africa: palaeocurrent-mediated chameleon dispersal to the Seychelles islands. *Biology Letters* 7 (2):225-228.

Ubick, D., and C. E. Griswold. 2011. The Malagasy goblin spiders of the new genus *Malagiella* (Araneae, Oonopidae). *Bulletin of the American Museum of Natural History* (356):1-86.

Upchurch, P. 2008. Gondwanan break-up: legacies of a lost world? *Trends in Ecology & Evolution* 23 (4):229-236.

Van der Meijden, Arie, Renaud Boistel, Justin Gerlach, Annemarie Ohler, Miguel Vences, and Axel Meyer. 2007. Molecular phylogenetic evidence for paraphyly of the genus *Sooglossus*, with the description of a new genus of Seychellean frogs. *Biological Journal of the Linnean Society* 91 (3):347-359.

Vaughan, JH. 1929. The birds of Zanzibar and Pemba. *Ibis* 71:577-608.

Vences, M, J Freyhof, R Sonnenberg, J Kosuch, and M Veith. 2001. Reconciling fossils and molecules: Cenozoic divergence of cichlid fishes and the biogeography of Madagascar. *Journal of Biogeography* 28:1091-1099.

Vences, M, J Kosuch, MO Rodel, S Lotters, A Channing, F Glaw, and W Bohme. 2004. Phylogeography of Ptychadena mascareniensis suggests transoceanic dispersal in a widespread African-Malagasy frog lineage. *Journal of Biogeography* 31:593-601.

Vences, M., K. C. Wollenberg, D. R. Vieites, and D. C. Lees. 2009. Madagascar as a model region of species diversification. *Trends in Ecology & Evolution* 24 (8):456-465.

Venkatasamy, S., G. Khittoo, P. Nowbuth, and D. R. Vencatasamy. 2006. Phylogenetic relationships based on morphology among the *Diospyros* (Ebenaceae) species endemic to the Mascarene Islands. *Botanical Journal of the Linnean Society* 150 (3):307-313.

Verneau, O., L. Du Preez, and M. Badets. 2009. Lessons from parasitic flatworms about evolution and historical biogeography of their vertebrate hosts. *Comptes Rendus Biologies* 332 (2-3):149-158.

Verneau, O., L. H. Du Preez, V. Laurent, L. Raharivololoniaina, F. Glaw, and M. Vences. 2009. The double odyssey of Madagascan polystome flatworms leads to new insights on the origins of their amphibian hosts. *Proceedings of the Royal Society B-Biological Sciences* 276 (1662):1575-1583.

Vidal, N., J. Marin, M. Morini, S. Donnellan, W. R. Branch, R. Thomas, M. Vences, A. Wynn, C. Cruaud, and S. B. Hedges. 2010. Blindsnake evolutionary tree reveals long history on Gondwana. *Biology Letters* 6 (4):558-561.

Vilgalys, R., and B. L. Sun. 1994. Ancient and recent patterns of geographic speciation in the oyster mushroom *Pleurotus* revealed by phylogenetic alanysis of ribosomal DNA-sequences. *Proceedings of the National Academy of Sciences of the United States of America* 91 (10):4599-4603.

Wallace, A. R. 1876. *The Geographical Distribution of Animals: With a Study of the Relations of Living and Extinct Faunas as Elucidating the Past Changes of the Earth's Surface*: Harper and brothers.

Warren, B. H., E. Bermingham, R. C. K. Bowie, R. P. Prys-Jones, and C. Thebaud. 2003. Molecular phylogeography reveals island colonization history and diversification of western Indian Ocean sunbirds (Nectarinia : Nectariniidae). *Molecular Phylogenetics and Evolution* 29 (1):67-85.

Warren, B. H., E. Bermingham, R. P. Prys-Jones, and C. Thebaud. 2005. Tracking island colonization history and phenotypic shifts in Indian Ocean bulbuls (*Hypsipetes*: Pycnonotidae). *Biological Journal of the Linnean Society* 85 (3):271-287.

Warren, B. H., E. Bermingham, R. P. Prys-Jones, and C. Thebaud. 2006. Immigration, species radiation and extinction in a highly diverse songbird lineage: white-eyes on Indian Ocean islands. *Molecular Ecology* 15 (12):3769-3786.

Warren, B. H., D. Strasberg, J. H. Bruggemann, R. P. Prys-Jones, and C. Thébaud. 2010. Why does the biota of the Madagascar region have such a strong Asiatic flavour? Cladistics 26: 526-538.

Waters, J. M. 2008. Driven by the West Wind Drift? A synthesis of southern temperate marine biogeography, with new directions for dispersalism. *Journal of Biogeography* 35 (3):417-427.

Weeks, A., D. C. Daly, and B. B. Simpson. 2005. The phylogenetic history and biogeography of the frankincense and myrrh family (Burseraceae) based on nuclear and chloroplast sequence data. *Molecular Phylogenetics and Evolution* 35 (1):85-101.

Wegener, A. 1912. Die Herausbildung der Grossformen der Erdrinde (Kontinente und Ozeane), auf geophysikalischer Grundlage. *Petermanns Geographische Mitteilungen* 63:185-195, 253-256, 305-309.

Wegener, A. 1966. *The origin of continents and oceans*. Translated by B. John: Courier Dover.

Wesener, T., M. J. Raupach, and P. Sierwald. 2010. The origins of the giant pill-millipedes from Madagascar (Diplopoda: Sphaerotheriida: Arthrosphaeridae). *Molecular Phylogenetics and Evolution* 57 (3):1184-1193.

Wesener, T., and D. VandenSpiegel. 2009. A first phylogenetic analysis of Giant Pill-Millipedes (Diplopoda: Sphaerotheriida), a new model Gondwanan taxon, with special emphasis on island gigantism. *Cladistics* 25 (6):545-573.

Weyeneth, N., S. M. Goodman, B. Appleton, R. Wood, and M. Ruedi. 2011. Wings or winds: inferring bat migration in a stepping-stone archipelago. *Journal of Evolutionary Biology* 24 (6):1298-1306.

Weyeneth, N., S. M. Goodman, W. T. Stanley, and M. Ruedi. 2008. The biogeography of *Miniopterus* bats (Chiroptera: Miniopteridae) from the Comoro Archipelago inferred from mitochondrial DNA. *Molecular Ecology* 17 (24):5205-5219.

Wikstrom, N., M. Avino, S. G. Razafimandimbison, and B. Bremer. 2010. Historical biogeography of the coffee family (Rubiaceae, Gentianales) in Madagascar: case studies from the tribes Knoxieae, Naucleeae, Paederieae and Vanguerieae. *Journal of Biogeography* 37 (6):1094-1113.

Wildman, D. E., M. Uddin, J. C. Opazo, G. Liu, V. Lefort, S. Guindon, O. Gascuel, L. I. Grossman, R. Romero, and M. Goodman. 2007. Genomics, biogeography, and the diversification of placental mammals. *Proceedings of the National Academy of Sciences of the United States of America* 104 (36):14395-14400.

Witt, J. D. S., R. J. Zemlak, and E. B. Taylor. 2011. Phylogeography and the origins of range disjunctions in a north temperate fish, the pygmy whitefish (*Prosopium coulterii*), inferred from mitochondrial and nuclear DNA sequence analysis. *Journal of Biogeography* 38 (8):1557-1569.

Wood, H. 2008. A revision of the assassin spiders of the *Eriauchenius gracilicollis* group, a clade of spiders endemic to Madagascar (Araneae : Archaeidae). *Zoological Journal of the Linnean Society* 152 (2):255-296.

Wood, H. M., C. E. Griswold, and G. S. Spicer. 2007. Phylogenetic relationships within an endemic group of Malagasy 'assassin spiders' (Araneae, Archaeidae): ancestral character reconstruction, convergent evolution and biogeography. *Molecular Phylogenetics and Evolution* 45 (2):612-619.

Yamagishi, S., M. Honda, K. Eguchi, and R. Thorstrom. 2001. Extreme endemic radiation of the Malagasy vangas (Aves : Passeriformes). *Journal of Molecular Evolution* 53 (1):39-46.

Yassin, A., L. O. Araripe, P. Capy, J. L. Da Lage, L. B. Klaczko, C. Maisonhaute, D. Ogereau, and J. R. David. 2008. Grafting the molecular phylogenetic tree with morphological branches to reconstruct the evolutionary history of the genus *Zaprionus* (Diptera : Drosophilidae). *Molecular Phylogenetics and Evolution* 47 (3):903-915.

Yoder, A. D., and M. D. Nowak. 2006. Has vicariance or dispersal been the predominant biogeographic force in Madagascar? Only time will tell. *Annual Review of Ecology Evolution and Systematics* 37:405-431.

Zhan, A. B., and J. Z. Fu. 2011. Past and present: Phylogeography of the *Bufo gargarizans* species complex inferred from multi-loci allele sequence and frequency data. *Molecular Phylogenetics and Evolution* 61 (1):136-148.

Zhang, P., and M. H. Wake. 2009. A mitogenomic perspective on the phylogeny and biogeography of living caecilians (Amphibia: Gymnophiona). *Molecular Phylogenetics and Evolution* 53 (2):479-491.

5

Hybridisation, Introgression and Phylogeography of Icelandic Birch

Kesara Anamthawat-Jónsson
Institute of Life and Environmental Sciences, University of Iceland,
Askja, Reykjavik,
Iceland

1. Introduction

Birch woodland is an integral component of the tundra biome, which covers expansive areas of the Arctic and amounts to 20% of Earth's land surface. Arctic tundra is located in the northern hemisphere, encircling the North Pole and extending south to the coniferous forests of the taiga. In geographical Europe, this includes, from east to west, northern Russia, Fennoscandia (northern Scandinavia and Finland), the Svalbard archipelago and Iceland, which is the main focus of the present paper. Tundra is the coldest of all of the terrestrial biomes and is noted for its frost-moulded landscapes, extremely low temperatures, little precipitation, poor nutrients and short growing seasons with long days (Reece *et al.* 2011). The low temperatures in the Arctic are, however, highly variable from one area to another, due to the influence of different oceanic currents. For example, the Irminger Current, which branches from the North Atlantic Current at about 26°W (Bersch *et al.* 1999), transports warm (4-6 °C) water around the coast of Iceland, making the climate in Iceland more temperate than its far north location would suggest.

Tundra vegetation is mostly herbaceous, consisting of a mixture of mosses and lichens, grasses and forbs, along with some dwarf shrubs and trees. The vegetation structure of tundra is simple, with low species diversity, and among the shrubs and trees, birch (*Betula* L.) is a dominant woodland plant (**Fig. 1**). There are no deep root systems in the vegetation of the Arctic tundra; however, a number of plant species are able to tolerate the cold climates. These plants are adapted to sweeping winds and soil disturbance, and can carry out photosynthesis at low temperatures and low light intensities. Plants are low-growing and group together, and thus can tolerate cold temperatures and are protected by snow during the winter. As the growing seasons are short, most of these plants reproduce more vegetatively than sexually by flowering. The plants, especially pioneering species, are generally wind-pollinated and have acquired an effective means of dispersal in the open landscape of the Arctic tundra. Birch has all of the above-mentioned characteristics, and in addition it can disperse effectively with its light, winged seeds.

Birches are pioneer tree species which, with their tiny wind-blown seeds, often rapidly colonize open areas, heathlands and marginal and disturbed habitats, such as forest clearings. Birches establish most effectively on bare soils; even in the lowest vegetation they

Fig. 1. *Betula pubescens* in the woodland Brekkuskógur, south-western Iceland, showing its typical shrub-like feature and autumn colour. The cold stream in the background originates from under a lava field. Photograph taken by KAJ.

grow very poorly (Kinnaird 1974; Aradóttir 1991), presumably due to the lack of affinity for any particular soil type, their ability to grow on nutrient-poor soils and their intolerance of shade. This pioneering birch establishment begins a process of succession which eventually converts the area into woodland, provided there is no outside intervention in the form of grazing or human activities. Birch was one of the first tree species to become established in northern Europe after the last Ice Age retreated. It is also one of the dominant woodland trees in the more extreme climate of the Arctic tundra. Birch migrated rapidly after the Last Glacial Maximum (LGM) from southern and central Europe and quickly colonized northern Europe (Bennett et al. 1991; Paus 1995; Willis et al. 2000). However, it was not until the beginning of the Holocene that birch pollen appeared in lake sediments and peat in Iceland (10.2–9.6 cal ka BP) and birch woodland only became established from about 8.5 cal ka BP (Hallsdóttir 1995; Karlsdóttir et al. 2009). Birch arrived in Greenland a few thousand years later (Fredskild 1991).

The first objective of this paper is to review palynological and molecular evidence supporting the postglacial origin of Icelandic birch and its phylogeographical patterns, both within Iceland and in relation to Europe. Most of the studies have been carried out in my research group (e.g. Karlsdóttir et al. 2009; Thórsson 2008; Thórsson et al. 2010). Aspects discussed in this paper include when and how birch came to colonize Iceland, the history of birch vegetation in Iceland after the first colonization, birch woodland expansion and regression through time, climatic and geographical structures that may influence woodland iability, as well as genetic and evolutionary factors that may have played critical roles in haping the birch woodlands seen today.

he birch species currently found in Iceland are extremely variable, both morphologically nd genetically. My own work, together with that of my research group during the past wenty years, has shown that introgressive hybridisation (introgression and gene flow) is robably the most significant drive towards the present-day variability in Icelandic birch. he second objective of this paper is therefore to review all evidence supporting ybridisation and introgression in *Betula*, notably interspecific hybrids from crossing xperiments (Anamthawat-Jónsson & Tómasson 1990 and 1999), a qualitative and uantitative assessment of morphological variation of birch in natural woodlands (Thórsson t al. 2001 and 2007), triploid birch hybrids (Anamthawat-Jónsson & Thórsson 2003; .namthawat-Jónsson *et al.* 2010) and birch palynology and Holocene hybridisation Karlsdóttir *et al.* 2007, 2008 and 2009). All of this evidence indicates that Iceland could be onsidered as a birch hybrid zone, harbouring genetic variation which is likely to be dvantageous in the arctic and subarctic environments. The present review provides an nsight into the introgression and phylogeography of Icelandic birch which should lead to a etter understanding of *Betula* in its broader geographical range, together with the bio- and •hylogeography of plant species on oceanic islands (especially in the North Atlantic region) nd the vegetation ecology and biodiversity conservation of the tundra biome.

:. Birch (*Betula* L.)

!etula is a genus of about 35–50 species distributed throughout the temperate, boreal and .rctic regions of the Northern Hemisphere. There is no consensus on species limits in *Betula*, vith different authors differing widely in what species they accept, from under 30 species to •ver 60. This is believed to be due to the fact that birches hybridise freely both in cultivation nd in nature, resulting in continuous variation in morphology among the species involved, ind hence making species delineation difficult.

\ccording to Flora Europeae, only four *Betula* species are recognized for Europe (Walters 964): two tree birch species, i.e. *B. pendula* Roth (silver birch) and *B. pubescens* Ehrh. (downy •irch), and two small shrub birch species, *B. humulis* Schrank (shrub birch of Central and :astern Europe) and *B. nana* L. (dwarf arctic birch); however, numerous other species •resent in Europe are considered to be conspecific to the above-mentioned species or are reated as subspecies, geographical variants and hybrids. Flora of North America recognizes .8 mostly native species of North America and the neighbouring northern regions (Furlow 997). North American birches are considered to belong to three groups: (1) the Costatae ;roup that consists of large trees, often with dark bark, such as the valuable timber tree ipecies *B. alleghaniensis* Britt., (2) the circumboreal Betula group that consists of small and nedium-sized trees, often with white bark, such as the paper birch *B. papyrifera* Marsh., and 3) the Nanae group of dwarf shrubby birches of the cold circumpolar region, such as the American dwarf birch *B. glandulosa* Michx.

:lora of China (Li & Cheng 1979; Li & Skvortsov 1999), on the other hand, has estimated a total)f 50–60 species in the genus *Betula*. The flora describes 32 mostly Asiatic species, 14 of which ire considered endemic to China. The geographical range covered by this flora is both vast ind diverse. For example, *B. ermanii* Chamisso, which is a highly valuable hardwood tree, can be found from the tundra of Kamchatka Peninsula in the Russian Far East to Japan and Korea and more inland into Mongolia. *Betula utilis* D. Don, a valuable timber tree of commercial importance commonly found in temperate broad-leaved forests at high altitudes (2500–3800 m), has its distribution range from Inner Mongolia north of China to Yunnan province in the south and over the Himalayan region of Afghanistan, Bhutan, India and Nepal. The distribution of *B. chinensis* Maximovicz, on the other hand, is more limited to the broad-leaved forests in mountain valleys and rocky mountain slopes in the northern part of China and Korea. This species is one of the most valuable timber trees in North China. *Betula alnoides* Buchanan-Hamilton ex D. Don, one of the tallest birch tree species, has its distribution in the subtropical forests that range from Central China to South Yunnan province and further south in the montane forests of Bhutan, Myanmar, Thailand and Vietnam. *Betula utilis* and *B. alnoides* are among the most ecologically important broad-leaved tree species along the Himalayan range (Zobel & Singh 1997; Gardner *et al.* 2000). Although some of the Asiatic species may no have justifiable species status, due to being conspecific, subspecies or hybrids, there are others that have yet to be discovered, especially those in remote and inaccessible areas. For example *Betula fujianensis*, a new species from subtropical evergreen and deciduous mixed forest in south-eastern China, has just recently been described (Zeng *et al.* 2008).

Birch (*Betula*) is a genus of monoecious trees or shrubs – a plant produces separate male and female catkins. *Betula* differs from its closely related genus *Alnus* Miller (alder) mainly in that the female (fruiting) catkins of birch are usually cylindrical in shape and the seeds have 3-lobed scales and fall with the fruit, whereas the female catkins of alder are ovoid in shape, woody and do not disintegrate at maturity, opening to release the seeds in a similar manner to that of many conifer cones (Walters 1964; Furlow 1997). Birch trees are not long-lived, rarely exceeding 80 years old. Birch woods may have a very diverse invertebrate life as the trees can support over 300 different species of insects and mites (Atkinson 1992; de Groot *et al.* 1997). This in turn attracts a variety of birds. Birch also has a number of fungi associated with it, especially beneficial soil mycorrhizal fungi that help to make the birch plants healthier and more resistant to insect herbivory, both above and below ground (Enkhtuya *et al.* 2003; Oddsdóttir *et al.* 2010). Birch is clearly vital to the viability and productivity of the tundra and forest ecosystems.

Birch is both ecologically and economically important. As can be seen in **Fig. 2**, research on *Betula* during the past 5–6 years has mostly been in the area of ecology and silviculture (261/339 publications or 77%). Examples of ecological papers include birch community and ecosystem analysis (Kleczewski *et al.* 2010; Deslippe *et al.* 2011; Takahashi *et al.* 2011), regeneration, vegetation succession and human influence (Shrestha *et al.* 2007; Erlendsson & Edwards 2009; Schofield & Edwards 2011), impact of climate change on plant performance and productivity (Levanic & Eggertsson 2008; Pudas *et al.* 2008; Sano *et al.* 2010), modelling of plant structure, growth and phenology (Linkosalo *et al.* 2010; Caffarra *et al.* 2011; Lintunen *et al.* 2011), environmental pollution and chemical ecology (Kontunen-Soppela *et al.* 2010; Franiel & Babczynska 2011; Morales *et al.* 2011). Although the category ecology-physiology appears to be the most published subject, a good proportion of the studies have direct application in silviculture and forestry.

Research on silviculture involves all aspects of establishment, growth, regeneration, composition, viability and quality of forests to meet diverse needs and values. Birch wood is

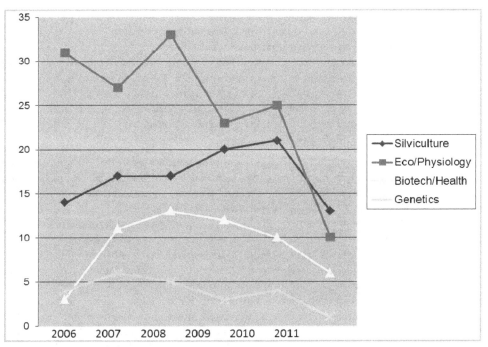

Fig. 2. Numbers and categories of publications revealed by a search in Web of Science using one keyword, Betula in Title, from 2006 to present. Note that 2011 is only up to September. The search found 339 publications, which were sorted manually into categories relevant to birch research after reading the content of the papers or the abstracts. Although the search is narrow, it shows a similar trend every year.

sought after for its fine quality, appearance, light colour, grain patterns, strength and durability. Birch is ranked among the highest quality timber for making furniture, in flooring and wall structure, and it also makes excellent plywood. There is great variation in the wood type and quality of different birch species, as there are diverse end-uses of birch wood. However, species that grow to medium- and large-sized trees (15–35 m in height) are the ones most used for commercial purposes. In Europe, silver birch (*B. pendula*) and downy birch (*B. pubescens*) probably produce the most valuable hardwood timber for the northern regions of Europe, including Scandinavia, Finland, Russia and the Baltic countries. Recent studies on these species include modelling and analysis of the growth and structure of forests and plantations (Hynynen *et al.* 2010; Kund *et al.* 2010; Lintunen *et al.* 2011); measuring and managing nutrient availability, retranslocation and fertilization (Mandre *et al.* 2010; Ruuhola *et al.* 2011); regeneration and seedling recruitment (Luostarinen *et al.* 2009; Sanz *et al.* 2011). In North America, paper birch (*B. papyrifera*), the medium-sized and fast-growing tree that forms pure stands or is otherwise found in mixed hardwood-conifer forests, is the most widely distributed birch species, especially in Canada. Its timber is used commercially for veneer, pulpwood and many speciality items. Recent studies on this birch are focused on timber quality in relation to tree growth, browsing, soil fertility and other factors (Droulin *et al.* 2010; Kleczewski *et al.* 2010; Belleville *et al.* 2011; Nielsen *et al.* 2011; Rea 2011). Asiatic and subtropical birches, on the other hand, receive very little attention regarding silviculture, possibly due to the abundance and diversity of hardwood species that may be more valuable. Nevertheless, Asiatic species such as the northern species *B. ermanii* are important hardwoods for construction and furniture-making, and strong wooden tools can be made out of the Chinese species *B. chinensis*; however, research on silviculture of these species is still limited (Tabata *et al.* 2010). Furthermore, birch is one of several subtropical and tropical tree species that are now protected by local conservation laws and difficult access, for example the Himalayan birch *B. utilis* and the Southeast Asian birch *B. alnoides*.

Interestingly, 16% of publications on *Betula* during the past few years (**Fig. 2**) come under the category biotechnology and health. There are two main issues in this category. Firstly, extracts and glycosides from birch leaves, twigs and bark have been tested in biochemical, molecular and cellular experiments to show medicinal or pharmaceutical properties. For example, extracts from *B. pendula*, *B. pubescens* and *B. platyphylla* have shown anti-inflammatory effects, anti-proliferation of human cells and improved immune response *in vitro* (Freysdottir *et al.* 2011; Grundemann *et al.* 2011; Huh *et al.* 2011). A number of new biologically active compounds have recently been isolated, especially from Asiatic birch (Phan *et al.* 2011; Xiong *et al.* 2011). Secondly, birch pollen causes allergy. In northern latitudes, birch is considered to be the most important allergenic tree pollen, with an estimated 15–20% of hay fever sufferers sensitive to birch pollen grains. Modelling and prediction of the duration and intensity of birch flowering, measurement of pollen accumulation rates and the biochemical identification of pollen antigens are among the most recent studies on *Betula* (Kuopparmaa *et al.* 2009; Linkosalo *et al.* 2010; Erler *et al.* 2011).

On the other hand, publications on *Betula* genetics, phylogenetics, vegetation history, phylogeography, genecology and related fields are surprisingly limited (**Fig. 2**), even though birch is an ecologically and economically important plant. Research studies in this category are very diverse: from mapping the demographic variation of dwarf birch *B. nana* (Ejankowski 2010) and developing molecular markers for use in the breeding of tree-birch species such as *B. pendula* and *B. alnoides* (Guo *et al.* 2008; Jiang *et al.* 2011) to resolving phylogenetic relationships among *Betula* species using genome-wide markers, nuclear genes or DNA barcodes (Li *et al.* 2007; Schenk *et al.* 2008; Crautlein *et al.* 2011) and answering questions about the origin, hybridisation, introgression and phylogeography of a number of *Betula* species using molecular markers, botanical and statistical approaches, and macro- and microfossil evidence (Nagamitsu *et al.* 2006; Maliouchenko *et al.* 2007; Thórsson *et al.* 2007; Truong *et al.* 2007; Karlsdóttir *et al.* 2009; Anamthawat-Jónsson *et al.* 2010; Thórsson *et al.* 2010; Tsuda & Ide 2010). It is indeed our studies of *Betula* species from Iceland that have made a major contribution to a better understanding of birch hybridisation, introgression and phylogeography in general. The following sections in this review are therefore about the studies of Icelandic birch species.

3. Introgressive hybridisation as a major player in the maintenance of genetic variation in Icelandic birch

Two species of *Betula* co-exist in Iceland: the dwarf birch *B. nana* and the downy birch *B. pubescens* (Stefánsson 1901; Gröntved 1942; Löve & Löve 1956; Thórsson *et al.* 2001). *Betula nana* is a prostrate shrub up to one metre in height, whereas *B. pubescens* may grow up to many metres tall. However, in Icelandic woodlands this tree birch species is often a shrub or low tree (**Fig. 1**). Both species are found together in most areas, although *B. pubescens* occupies lower elevations and a drier habitat compared with *B. nana*. In the forestry context, birch (i.e. *B. pubescens*) is the only tree-forming natural woodland in Iceland. Birch woodland in Iceland covers only about 1% of the total land area today, but birch is believed to have had an almost continuous distribution, covering most of Iceland's lowlands before the first settlement in the ninth century (Hallsdóttir 1995; Kristinsson 1995). During the last centuries, deforestation has been a continuous process, resulting in a highly fragmented distribution of birch populations.

Taxonomically, *B. nana* is represented by subspecies *nana* (Suk.) Hultén in Europe and western Asia, and by subspecies *exilis* (Suk.) Hultén in North America and central and eastern Asia (Hultén & Fries 1986). Only the subspecies *nana* is found in Iceland. *Betula pubescens* is a European species, represented by subspecies *pubescens* Ehrh., which may grow up to 25 m tall with single (monocormic) or many (polycormic) stems, and by subspecies *ortuosa* (Ledeb.) Nyman, which is a shrub or low tree found in the mountain regions of northern Europe (Walters 1964). The latter subspecies, so-called mountain birch, is believed to be the result of introgressive hybridisation with *B. nana* (Vaarama & Valanne 1973; Kallio *et al.* 1983); the same process that occurs with Icelandic birch (Anamthawat-Jónsson 1994 and 2003). In Iceland, *B. pubescens* tends to be in the form of 1–2 m low shrubs, especially in the regions with extreme oceanic climate and heavy coastal storms. Along the tree line in the mountains a zone of birch shrubs growing horizontally on the ground can be seen throughout the country. Due to the extensive and continuous morphological variation of birch in Iceland, the tree birch species is not divided into subspecies, but is treated in this review as *Betula pubescens* sensu lato.

Morphological variation in birch (*Betula*) is known to be extensive, due in part to frequent hybridisation in this genus (Woodworth 1929; Johnsson 1945; Walters 1964; Furlow 1997). This has made taxonomy of *Betula* problematic. Because of the difficulty in delineating species, morphological criteria for identifying interspecific hybridisation or hybrids can be extremely ambiguous. In the areas where the distribution of birch species overlaps, plants with intermediate morphology (presumed hybrids) have frequently been noted (reviewed in Atkinson 1992). Some of these putative hybrids have turned out to be part of the large range of variability within species. But hybrids between species with different ploidy levels can be confirmed cytogenetically. Such is the case with birch hybrids found in Iceland. Hybridisation between diploid *B. nana* and tetraploid *B. pubescens* produces triploid hybrids and this can be unequivocally confirmed by counting the chromosomes in somatic metaphase cells of the plants (**Fig. 3**). Triploid chromosome number has been confirmed among hybrids between *B. nana* and *B. pubescens* in crosses (Anamthawat-Jónsson & Tómasson 1990 and 1999) and among plants in natural woodlands (Thórsson *et al.* 2001 and 2007).

Triploid chromosome number indicating natural hybridisation between diploid and tetraploid birch species in Europe has been reported, but most of the plants are thought to be putative hybrids between silver birch (diploid *B. pendula*) and the tetraploid *B. pubescens* (Brown *et al.* 1982; Brown & William 1984). Natural triploid hybrids between *B. nana* and *B. pubescens* might have been among the plants with intermediate morphology found in Scotland (Kenworthy *et al.* 1972) and Fennoscandia (M. Sulkinoja, pers. comm.), but there was no cytogenetic confirmation. Birch chromosomes are extremely small and numerous,

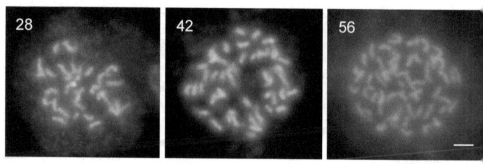

Fig. 3. Chromosome number of diploid Betula nana (2n = 2x = 28), triploid hybrid (2n = 3x = 42) and tetraploid B. pubescens (2n = 4x = 56). The chromosomes were isolated from leaf buds by Æ. Th. Thórsson, following the protocol of Anamthawat-Jónsson (2004), and stained with the fluorochrome DAPI. Scale bar represents 2 µm.

thus there is often uncertainty in counts from individual cells. Metaphase chromosomes of B. pendula (diploid) and four American tetraploid birch species are only 0.6–1.8 µm long (Taper & Grant 1973). Most of the B. pubescens chromosomes are less than one µm long (Anamthawat-Jónsson & Heslop-Harrison 1995). Birch genomes are also very small. Our recent investigation of Icelandic birch species (Anamthawat-Jónsson et al. 2010) shows that the average 1C genome size of the tetraploid B. pubescens (882 Mbp, 0.90 pg) is about twice the size of the genome of the diploid species B. nana (448 Mbp, 0.46 pg), while the triploid group has an average size of 666 Mbp (0.68 pg), which is mid-way between the two species. For comparison, the genome size of a diploid (2n = 2x = 14) barley Hordeum vulgare is 5.55 pg and that of a hexaploid (2n = 6x = 42) wheat Triticum aestivum is 17.33 pg (Bennett & Smith 1976). The latest compilation of angiosperm genome sizes (Bennett & Leitch 2011) includes new record holders for the smallest (1C = 0.0648 pg in Genlisea margaretae) and largest (1C = 152.23 pg in Paris japonica) genome sizes so far reported.

As stated previously, birch chromosomes are numerous and minute. Furthermore, the conventional squash method of root-tip chromosome preparation is often ineffective because rooting of birch is very difficult. But by using the protoplast dropping method of chromosome preparation developed for Betula (Anamthawat-Jónsson 2004), it is possible to obtain high-quality metaphases from young leaf buds collected in the field at any time of the growing season. This has made chromosome counting accurate and cytogenetic investigation of birches at a population level possible.

By having the means to accurately identify triploidy among birch plants in the field, we are able to investigate the occurrence of natural hybrids in Iceland where two birch species, B. nana and B. pubescens, coexist and the growing season is short enough for their flowering to be relatively synchronous. Hybridisation is known to be an important factor that influences evolution in a variety of ways (Barton 2001); for example many plant species have a hybrid ancestry (Rieseberg 1997; Rieseberg & Willis 2007), including Betula (Nagamitzu et al. 2006). The most common form of hybrid speciation is probably allopolyploidy, whereby interspecific or intergeneric hybridisation gives rise to new species following chromosome doubling and reproductive isolation from the parental species. Natural hybridisation can also lead to gene flow between species through backcrossing of the hybrid with its parental species; this process is known as hybrid introgression, introgressive hybridisation, or simply introgression (Heiser 1973). Such is the case with birch hybridisation described in this review. Only a few partially fertile hybrids are sufficient for introgression to occur, and therefore these hybrids may not have been detected. Introgressive hybridisation allows for the transfer of neutral or adaptive traits from one species to another and can increase genetic polymorphism in one or both parental species, but it may have a negative outcome, such as an evolution of aggressive weeds, or result in the extinction of a species. Numerous cases of introgressive hybridisation have been documented, including plants in natural habitats and under cultivation (e.g. Arnold 1992; Ellstrand et al. 1999; Jarvis & Hodgkins 1999; Minder & Widmer 2008).

Introgression in birch was examined originally using morphological characters, in particular introgression between B. nana and B. pubescens in Iceland, northern Scandinavia and Greenland (e.g. Elkington 1968; Vaarama & Valanne 1973; Sulkinoja 1990). Most studies described introgressive modification of the tree-birch species as a distinct type, which became recognized as mountain birch (Betula pubescens ssp. tortuosa). But our studies using botanical, cytogenetic and molecular approaches revealed that the introgression in birch is bi-directional, resulting in gene flow between the two species via triploid interspecific hybrids (Anamthawat-Jónsson & Thórsson 2003; Thórsson et al. 2001, 2007 and 2010). Such gene flow is likely to be an important mechanism in maintaining genetic variation in both species.

We have examined the morphological variation systematically; the results are summarized here in **Fig. 4** (modified from Thórsson et al. 2007). The study includes more than 400 birch plants, randomly chosen regardless of morphology, from all major birch woodlands in Iceland. Chromosome counts from mitotic metaphases of all individuals under study produce three unambiguous groups of birch: diploid (2n = 2x = 28), triploid (2n = 3x = 42) and tetraploid (2n = 4x = 56). Of the 461 plants examined, 176 plants (38.2%) are diploid, 241 plants (52.3%) are tetraploid and 44 plants (9.5%) are triploid hybrids. The three ploidy groups are confirmed by genome size analysis on a subset of samples based on flow cytometry and DNA densitometry (Anamthawat-Jónsson et al. 2010). No aneuploid was found. Aneuploids have never been found among birch plants in nature, or from crosses. Triploid hybrids between B. nana and B. pubescens were backcrossed using B. pubescens as pollen donor – only triploids and tetraploids were discovered among the backcrossed progeny (Anamthawat-Jónsson & Tómasson 1990 and 1999). I therefore hypothesized that viable gametes produced by the triploid plants, although in extremely low frequencies, must have been "euploid" gametes with n = 14 or 28, most probably derived via a meiotic non-disjunction.

By looking at the plant morphology, i.e. analysing species-specific botanical characters, we can separate the diploid and the tetraploid groups most of the time (**Fig. 4**). The diploid group consists mostly of dwarf birch B. nana, whereas the tetraploid group predominantly includes B. pubescens-like plants. The triploid plants, most interestingly, resemble B. nana or show intermediate morphology. They rarely resemble B. pubescens. Morphological characters, especially in the leaf shape, are useful for species identification as they are relatively independent of environmental changes. Leaves of B. pubescens have been taxonomically described as being cordate with dentate margins, whereas B. nana leaves are orbicular with crenate margins. In order to define morphological variation within and across the species, a set of species-specific characters that can be visually scored has been

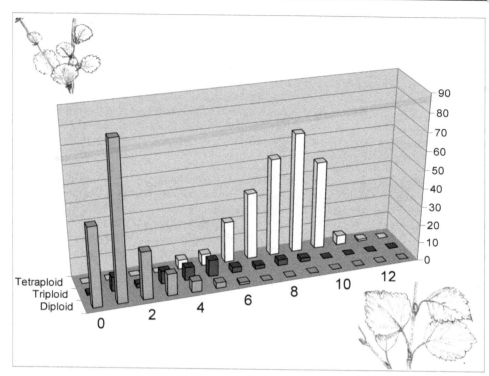

Fig. 4. Morphological distribution of 461 birch plants from 14 major woodlands throughout Iceland (modified from Thórsson 2008). Drawings by K. Anamthawat-Jónsson show a typical *Betula nana* plant (upper left corner) and *B. pubescens* (lower right corner). X-axis: Morphology indices from 0 (*B. nana*) to 13 (*B. pubescens*) based on species-specific botanical characters (see text). Y-axis: Number of plants belonging to each group by morphology index. The diploid, triploid and tetraploid groups had average scores of 1.3, 4.1 and 8.3 respectively. Introgression in Icelandic birch is evidently bi-directional.

developed, based on botanical characteristics (Clapham *et al.* 1962; Elkington 1968; Kenworthy *et al.* 1972), and was used in this study. The growth form and habit can be assessed in the field, but the leaf characters are usually scored from 30 randomly collected leaves per plant. Leaf shape characters are highly uniform within plants. The scores have been assigned to place *B. nana* at the lowest ranks (zero) and *B. pubescens* at the highest (13) as follows: shrub (0) or tree (1); growth habit procumbent (0) or erect (1); petiole sessile (0), intermediate (1) or non-sessile (2); leaf tip obtuse (0), sub-acute (1) or acute (2); leaf base rounded (0), cordate (1) or cuneate (2); leaf margin crenate (0), serrate (1) or dentate (2); leaf shape orbicular (0), obovate (1), or ovate (2); and leaf teeth single (0) or multiple (1). For each plant, the scores of all characters are combined into a single value called a morphology index, and this was plotted against ploidy of the plants.

Based on the qualitative analysis of species-specific characters (**Fig. 4**), diploid and tetraploid plants form separate peaks of morphology index distribution, representing *B. nana* and *B. pubescens* respectively. The *B. nana* peak includes the scores from 0 to 6 (average score 1.3). About 80% of diploid plants fall within the narrower morphology index range of 0-1 and only 23% of these have the minimum score zero, which is taxonomically equivalent to being pure *B. nana*. The scores 2-6 extend into the intermediate region of the total distribution, meaning that about 20% of the diploid plants look more like hybrids and much less like *B. nana*. At the high end of the morphology index, a broader peak of 4-12 scores (average 8.3) belongs to the tetraploid group, the *B. pubescens* group. About 83% of the tetraploid plants have most of the *B. pubescens* morphology (scores 7-10); about 15% are hybrid-like (scores 4-6) and could be mistaken for triploid birch; and only some 2% score higher than 10. Taxonomically, the maximum score for *B. pubescens* should be 13, but none of our tetraploid plants have this morphology. In other words, we have not found pure *B. pubescens* in natural birch woodlands in Iceland so far. Only one *B. pubescens* individual (out of 241 tetraploid plants) has a score of 12. Based on the average score of 8.3, most of the Icelandic tetraploid birch plants looked more like hybrids and much less like the typical European *B. pubescens*. And if we look at plants with intermediate morphology, with a score of 4 in particular, the plant can be diploid, triploid or tetraploid. Introgressive hybridisation is obviously evident among Icelandic birches.

The species-specific botanical characters of this group of 461 plants have also been measured quantitatively (Thórsson *et al.* 2007; Thórsson 2008), using the leaf morphology analysis program WinFolia (Regent Instruments, Quebec, Canada). Multivariate analysis of variance (MANOVA) was used to test the differentiation among ploidy groups and among sites within each group, whereas the linear discriminant analysis (LDA) was conducted to evaluate how the variables can be used to classify the different individuals. The principle of both methods is based on a linear combination of variables that maximizes the ratio of between-groups variance to within-groups variance (Quinn & Keough 2002). The homogeneity of variances of each variable was tested with the Bartlett test (Sokal & Rohlf 1995). The overall results strongly support the introgression study based on the morphology index described above. The linear discriminant analysis reveals significant separation among the three ploidy groups and the model assigned 96% and 97% of the *B. nana* and *B. pubescens* individuals correctly. The triploid hybrids are difficult to predict since only half of them could be assigned correctly. Among the species-specific leaf characters, leaf length is the most useful variable for identifying triploid hybrids.

Most interestingly, there is a clear indication of geographical structure among the woodlands investigated when the ploidy groups were analysed separately. The multivariate data analysis (Thórsson 2008) reveals geographical patterns within the ploidy groups, which could partly be explained by differences in mean July temperature. In the woodlands where summer is cold (often associated with glacial sites or the interior highlands), the leaf morphology in all ploidy groups tends to be closer to the minimum LDA values and has a low morphology index. On the other hand, the woodlands in lowland areas tend to have morphology closer to the maximum LDA values with a high morphology index. Shrub-like birch with intermediate or hybrid-like morphology is also known to be common in regions characterized by cold climates, such as Fennoscandia, the highland areas of Scandinavia, other mountain regions of Europe, and southern Greenland (e.g. Kallio *et al.* 1983; Gardiner 1984; Sulkinoja 1990; Jetlund 1994). In the northern part of the Urals and Western Siberia, in the region of forest tundra-taiga, changes in leaf parameters in *B. pubescens* including shape and complexity were found to correlate with climatic conditions such as long-term average temperatures (Migalina *et al.* 2010); this may have physiological advantages, especially in photosynthesis. Such morphological differentiation is likely to be driven by the introgressive hybridisation process, if the introgressant types are more adaptable (or more tolerant) to environmental pressure and habitats such as those found in Iceland and elsewhere in the subarctic regions. This introgressed birch can have certain advantages: for example the ability to spread vegetatively and form a large multicormic shrub could ensure survival of the plant in extreme environments. A molecular study on alpine sedge has shown that genotype integrity is maintained in optimal habitats, whereas introgressed individuals are favoured in marginal habitats (Choler *et al.* 2004). Our future work on birch introgression will most likely be in the area of experimental genecology.

4. Introgression, origin and molecular phylogeography of Icelandic birch

Introgressive hybridisation has been shown to be associated with demographic history of the species. For tree and woody species, a number of molecular studies have revealed dynamic patterns of demographic history in relation to the glacial histories of Europe, Asia and North America, as well as in the arctic–alpine regions (Taberlet *et al.* 1998; Petit *et al.* 2002; Skrede *et al.* 2006; Alsos *et al.* 2007; Fussi *et al.* 2010; Tsuda & Ide 2010; Seiki *et al.* 2011). These studies infer changes, including population expansion, plant migration and hybridisation, which lead either to allopolyploid speciation or introgression.

Icelandic birch samples from the morphological introgression study described above have been analysed molecularly, together with new samples collected in northern Scandinavia, Scotland and Greenland, amounting to 463 birch plants/trees in total (Thórsson 2008; Thórsson *et al.* 2010). The objective was to find direct genetic evidence supporting introgressive hybridisation between tetraploid tree birch (*Betula pubescens*) and diploid dwarf birch (*B. nana*) via triploid hybrids, and to investigate an association between the introgression and phylogeographical distribution of Icelandic birch. No genetic data was available on the origin and phylogeography of *Betula* in Iceland. In our study, chloroplast (cp) DNA haplotypes across different ploidy groups in two species of Icelandic birch were examined and the results were analysed in an attempt to relate the extent of introgression to the historical and current geographical distribution. Individual plants were first classified by their ploidy status after direct chromosome counting, according to the chromosome isolation protocol of Anamthawat-Jónsson (2004). The haplotypes were identified using the polymerase chain reaction-restriction fragment length polymorphism (PCR-RFLP) universal primers for non-coding regions of plant chloroplast genomes from Taberlet *et al.* (1991) and Demesure *et al.* (1995), following the protocol modified from Palmé *et al.* (2004). The geographical distribution of the haplotypes obtained was mapped, and subsequently the haplotype variation and introgression ratios (*IG*) were analysed statistically.

Thirteen haplotypes have been identified among 345 Icelandic samples (**Fig. 5**), along with five haplotypes from 118 samples from outside Iceland including three haplotypes shared with the Icelandic set of samples. These three are also the most common haplotypes in Iceland, with haplotypes T, C and A found in 49%, 19% and 15% of the plants respectively. Haplotypes D and F are represented by 4–6% of the plants, whereas other haplotypes (eight in total) are considered rare (less than 2%). All common haplotypes are shared between the triploid group and the parental species (**Fig. 5**), clearly indicating introgressive hybridisation.

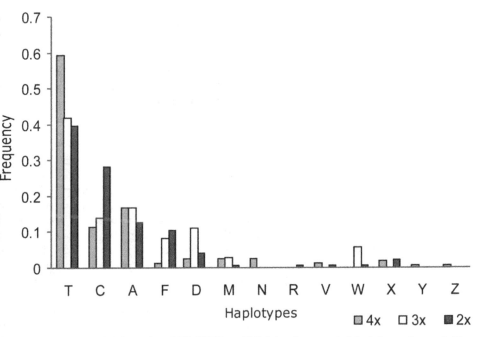

Fig. 5. Frequencies of chloroplast PCR-RFLP cp-DNA haplotypes in birch from three ploidy groups in Iceland (modified from Thórsson 2008): tetraploid (4x) *Betula pubescens*, diploid (2x) *B. nana* and triploid (3x) hybrid between the two birch species.

Introgressive hybridisation, as shown by the extensive sharing of common haplotypes across ploidy groups (**Fig. 5**), is supported by the statistical analysis of IG indices and the variation components. In this study, the observed and expected introgression ratios (IG and IG^e respectively) were calculated according to Belahbib et al. (2001), whereby IG reflects the amount of locally shared (within site) haplotypes between two species and IG^e represents the expected value if the haplotype sharing is not geographically structured. As gene flow among sites may be extensive, a modified ratio IG^* was also applied to contrast a species within a site with the other species pooled over all localities. Furthermore, modified ratios IGR and IGR^* (Palmé et al. 2004) were applied to this data set in an attempt to distinguish recent introgression from ancestral polymorphism (shared polymorphisms due to incomplete lineage sorting), although not as quantitatively. The results are as follows (see numerical results in Thórsson et al. 2010): (1) There is significant introgression in the Icelandic birch populations, which supports the haplotype sharing observed here and the morphological analysis described in the previous section. (2) There is indeed a geographical structure of introgression, for example the populations that are least introgressed are those woodlands in which summers are cold and often associated with glacial sites or the interior highlands, whereas the woodlands in optimal climatic environments harbour a great deal of gene flow. (3) The population differentiation appears to be more pronounced in *B. nana* than in *B. pubescens* with regard to introgression. (4) The overall introgression index for Iceland (all woodlands together) is nearly twice as large as the value for northern Scandinavia.

When the statistical analysis of *IG* indices and the dissection of haplotype variation components are combined, considerable differences are found to exist among samples in Iceland, shaped by isolation by distance and local introgression.

An east-west phylogeographical distribution in Iceland can be deduced from this molecular study. The overall geographical distribution of haplotypes in Iceland (see illustrations in Thórsson 2008) shows that the T haplotype (the most common haplotype, with 49% occurrence) is present in all woodlands examined, and often in all three ploidy groups. This haplotype is also prevalent in northern Scandinavia, where it is apparently associated with *B. nana*. Therefore the T haplotype might have been distributed over all woodlands and across ploidy groups via introgression and hybridisation. On the other hand, the distribution of all other haplotypes in Iceland is found to be associated with the geographical location of woodlands, i.e. forming an east-west separation pattern. Haplotype A (15% land occurrence) occurs mainly in western Iceland, especially in the north-west fjords where it is recorded at high frequency and in all ploidy groups. This haplotype is significantly correlated with longitudes. It is also a common haplotype among tree-birch species in Europe (Palmé *et al.* 2004) and is confirmed in the present study. Two rare haplotypes unique to Iceland (M and N) are found mainly in the western coastal woodlands. On the other hand, a greater diversity of haplotypes is observed from the eastern and north-eastern woodlands. Haplotype C (19% land occurrence, the second most common haplotype in Iceland after T), which also exists in all ploidy groups, is a dominant haplotype of the eastern woodlands. Interestingly, this is the most common birch haplotype in Europe (Palmé *et al.* 2004) and is also the most common haplotype among birch samples we collected from Northern Scandinavia and Scotland, yet it is essentially absent from Greenland. Of the ten haplotypes unique to Iceland, eight of them, including haplotypes D and F (4–6% occurrence) together with all other rare haplotypes, occur almost exclusively in the eastern sites. The higher number of rare haplotypes in the east could indicate that the eastern (and north-eastern) populations are more ancient in origin than the western (and south-western) woodlands. The east-west separation of haplotypes is also reflected clearly in the statistical analysis of phylogenetic association among haplotypes discovered in Iceland (Thórsson *et al.* 2010). The observed east-west haplotype distribution within Iceland may indicate different population histories or multiple origins of Icelandic birch.

Present-day birch in Iceland is most probably post-glacial in origin, i.e. having colonized Iceland in the early Holocene. The first colonization of Holocene birch is believed to have occurred in the north and north-eastern valleys and this is thought to be the result of different deglaciation patterns in that area, together with early-Holocene warming in northern Iceland. The Holocene vegetation history constructed from pollen records from lake sediments and macroscopic remains in peat (Hallsdóttir 1995; Hallsdóttir & Caseldine 2005) supports the hypothesis that birch woodland (dominated by the tree-birch species *B. pubescens*) started to form in the north-eastern valleys of Skagafjördur and Eyjafjördur in the Late Boreal and progressed into the western fjords and southern Iceland towards the end of the Atlantic period. We measured *Betula* pollen from an early Holocene peat profile from Hella in Eyjafjördur, mid-northern Iceland, with 39 samples taken at ca. 100-year intervals between ca. 10.3 and 7.0 cal ka BP based on known tephra layers, including the Saksunarvatn tephra at 10.2 cal ka BP (Karlsdóttir *et al.* 2009). The study shows two periods with large quantities of *B. pubescens* pollen (i.e. birch woodland expansion): the earlier period from approximately 9.3 to 8.3 cal ka BP and the later from approximately 7.3 cal ka BP to the end of the profile, with a peak at 7.2 cal ka BP. The early onset of vegetation development in northern Iceland is thought to be the result of different deglaciation patterns in that area (Norddahl 1991), together with early-Holocene warming in northern Iceland (Rundgren 1998; Caseldine et al. 2006). The Holocene Thermal Maximum (HTM) in terrestrial Iceland has been estimated between 10.3 and 5.6 cal ka BP (Kaufman et al. 2004) and the peak may have been about 7.1 cal ka BP, with the summer temperatures almost 1.5 °C above the averages of the 20th century (Wastl et al. 2001). The early Holocene peaks of B. pubescens pollen in northern Iceland, as represented by the Hella peat profile (Karlsdóttir et al. 2009), coincide with warming periods of the Boreal based on the Greenland ice-core project (Bond et al. 1997).

Our molecular studies seem to indicate that birch – particularly B. pubescens – colonized Iceland more than once during the early Holocene. The east-west separation of cpDNA haplotypes described above is so far the strongest indication of multiple origins in Icelandic birch. The higher number of rare haplotypes in the east could indicate that the eastern (and north-eastern) populations are more ancient in origin than the western (and south-western) woodlands. The two common haplotypes, i.e. A (west) and C (east), have clearly not crossed the geological boundary which includes the central highland and glaciers. The T haplotype, on the other hand, is distributed all over Iceland, which may be explained if this haplotype was B. nana-specific in the first place. Betula nana is an arctic-alpine species that colonized Iceland long before B. pubescens (Hallsdóttir 1995; Rundgren 1998), while new pollen records (L. Karlsdóttir, unpublished) indicate that B. nana established itself as early in the north (north-east) as in the south (south-west). Chloroplast genes, which are maternally inherited, could easily be transferred across species via hybrids, as B. nana is known to be a predominant seed parent in such hybridisation (Eriksson & Jonsson 1986; Anamthawat-Jónsson & Tómasson 1999).

Our current and ongoing work uses bi-parentally inherited nuclear DNA markers to further resolve the phylogeographical pattern of birch in Iceland. Tsuda & Ida (2010) found north-south separation of chloroplast DNA variation in Betula maximowicziana endemic to Japan, whereas the analysis using microsatellite (SSR) nuclear markers on this birch (Tsuda & Ide 2005) revealed dispersal by pollen across the boundary, at least in one direction. Genome-wide nuclear markers offer excellent within-species resolution, and are hence useful for dissecting genetic variation due to founder effects, genetic drift, bottleneck effects, gene flow and other evolutionary forces. Amplified fragment length polymorphic (AFLP) genome-wide markers have been used successfully to identify glacial survival of two west-arctic species, while the sequencing of non-coding regions of chloroplast DNA revealed only limited variation (Westergaard et al. 2011a).

The unique feature of our Betula research is that we have karyotyped plants in natural woodlands, meaning that our continuing molecular studies, for example using nuclear markers, can also answer specific questions about hybridisation and introgression. Once correlated to present-day and past biogeography, it will be possible to predict future changes due to human influences and global warming. We have already found evidence of hybridisation between B. nana and B. pubescens in Iceland as early as the first woodland establishment in the early Holocene (**Fig. 6**; Karlsdóttir et al. 2009), by using morphometric

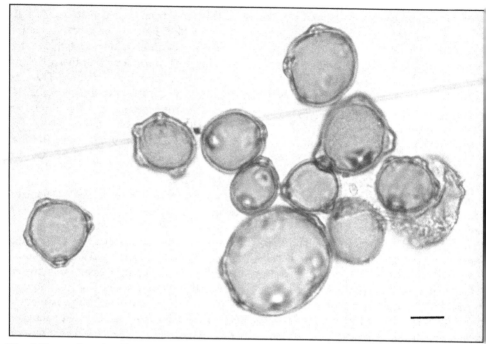

Fig. 6. Examples of pollen grains produced by triploid *Betula* hybrids (scale bar represents 10 μm). Triploid hybrids produce pollen with distinctive characteristics which may be used to detect hybridisation in sub-fossil samples. Several anomalies in pollen morphology, especially an unusual number of pores (4 or more), are significantly more frequent in pollen samples from triploid hybrids than from diploid and tetraploid species.

standards of pollen measurements obtained from karyotyped present-day birch plants. The frequency of non-triporate pollen (the strongest evidence of hybridisation) in the period between 9.2 and 8.7 cal ka BP, during the first birch woodland establishment in the north of Iceland, far exceeded the average level produced by the present-day triploid hybrids. Climatic and ecological conditions may have favoured hybridisation of birch species during the expansion of birch woodlands in warm periods.

One of the most important questions to us is where the Icelandic birch came from, and how. Our study of cpDNA variation (Thórsson *et al.* 2010) indicates that Icelandic birch is most likely European in origin and had colonized Iceland in the early Holocene. All three most common haplotypes found in Iceland (T, C and A haplotypes) are widespread in Europe (Palmé *et al.* 2004). Microsatellite data of Maliouchenko *et al.* (2007) also supports this similarity. Our study shows further that these cpDNA haplotypes are not prevalent at all sites investigated. Haplotype T, the most common haplotype in Iceland, dominates the northern Scandinavian sites and seems to be associated with *B. nana*, as suggested earlier in this section. Haplotype C (Icelandic eastern haplotype) is also found in Sweden and Scotland but is absent from Greenland. This haplotype is the most common and probably the oldest haplotype of birch in Europe, occurring in all three *Betula* species (i.e. *B. pendula*, *B. pubescens* and *B. nana*), and is most prevalent on the European mainland (Palmé *et al.*

2004). On the other hand, the A haplotype, which is prevalent among the western Icelandic populations, is also detected in birch samples from Greenland, Scotland and Scandinavia. It is the second most common haplotype of birch in Europe, occurring essentially in the tree-birch species, *B. pendula* and *B. pubescens*, and mainly in the western part of Europe, including the British Isles and Scandinavia (Palmé *et al.* 2004).

The most likely scenario could be that the first birch colonization (*B. nana* in particular) in the north and north-eastern part of Iceland came from Western Europe, and perhaps Scandinavia, to Iceland and eastern Greenland via the North Atlantic current. This is thought to be the most probable means of biota dispersal during the early Holocene (Buckland *et al.*, 1981). Bennike *et al.* (1999) carbon-dated plant and animal remains in sediments from north-eastern Greenland locations and concluded that woody plant species, including the dwarf birch *B. nana*, were Holocene immigrants and that the first immigrants came from north-west Europe. A number of molecular studies have shown that circumpolar species tend to split into Eurasian and North American lineages, and that the Eurasian lineage (the amphi-Atlantic species in particular) often has its distribution from Eurasia (western Siberia), northern Scandinavia and Iceland to eastern Greenland. Examples include *Juniperus* (Adams *et al.* 2003), *Saxifraga* (Abbott *et al.* 2000), *Vaccinium* (Alsos *et al.* 2005) and *Carex* (Westergaard *et al.* 2011b). An analysis of the history of the North Atlantic biota (Brochmann *et al.* 2003; Alsos *et al.* 2007) indicated that the majority of endemic species have undergone extensive migration and are post-glacial in origin.

Betula pubescens is also thought to have arrived in Iceland from Western Europe, but a few hundred years after *B. nana*. As shown earlier in this section, the first birch woodland establishment was in the northern (north-eastern) valleys. The eastern and north-eastern cpDNA haplotypes in Iceland seen today must have originated earlier than the western haplotypes, i.e. from the first dense woodlands in the Late Boreal, thus having enough time to evolve into the high haplotype diversity that contains several new and unique haplotypes seen today. On the other hand, the western and south-western haplotypes might have arrived much later, during the southern woodland establishment in the Atlantic period or during the regeneration of birch woodland in the lowlands of the south in the latter half of the Holocene. Pollen studies of early Holocene peat from south-western Iceland (e.g. Vasari 1972) appear to indicate fragmented or scattered woodlands in the area at the time. This is most likely due to a different geological history and palaeoenvironment during the early Holocene in South Iceland (Geirsdóttir *et al.* 2000), as a consequence of substantial sea-level changes during the Weichselian deglaciation (Norddahl & Pétursson 2005). The cpDNA haplotype diversity in western and south-western populations, which is only one-third of that found among the eastern and north-eastern woodlands, indicates more recent origin. But as the western haplotypes are essentially exclusive to western and south-western woodlands, there is only one explanation: birch in these woodlands originated from a different colonization, though also from Europe.

The migration of birch (*B. nana* and *B. pubescens*) to these regions could have come about easily by long-distance wind dispersal, as the winged seeds of birch are lightweight. The alternative route of Holocene dispersal, which is especially important for species lacking a wind-dispersal mechanism, is by ice-rafting and wood-drifting with the transpolar current from Siberia to northern Scandinavia, Iceland and eastern Greenland (Johansen & Hytteborn 2001). The arctic flora appears to be highly mobile (Alsos *et al.* 2007), as some dispersal vectors may be particularly efficient in the Arctic as a result of the open landscape, strong winds and extensive snow and ice cover.

Birch migrated rapidly after the Last Glacial Maximum (LGM) from southern and central Europe and quickly colonized northern Europe. Towards the end of the LGM, the arctic-alpine dwarf birch B. *nana* already covered much of central and eastern Eurasia (Tarasov *et al.* 2000). Based on macrofossil data (Fredskild 1991), dwarf birch occupied the North Atlantic coast of Scotland and south-western Norway around 11.2–10.3 cal ka BP. However, based on new macrofossil data (Birks & van Dinter 2010), B. *nana* had already arrived and spread by the Allerød period (13.5–12.9 cal. ka BP) at the Tjørna site, south-western Norway, then disappeared during the Younger Dryas but appeared again at the transition towards the Holocene, about 500 years before the arrival of B. *pubescens* in the area at around 11.4 cal ka BP. Note that the calibrated year BP is based on the early tephra horizons with Saksunarvatn tephra (Icelandic origin, ca. 10.2 cal ka BP), forming a horizon in the late Pre-Boreal of Northern Europe, and with Vedde ash (also Icelandic in origin, ca 12.0 cal ka BP). According to Paus (1995), B. *pubescens* existed in southern Norway as early as 10.8 cal ka BP. Both *Betula* species were found in the Shetland Islands in ca. 8.6 cal. ka BP (Fredskild 1991), but note that this estimation was based on macrofossil rather than pollen data. New evidence inferred from high-resolution plant macrofossil and pollen data (Hannon *et al.* 2010) showed that B. *nana* was already prevalent in the Faroe Islands during the earliest part of the Holocene (11.3–10.3 cal. yr BP). As shown by palynological studies (especially Karlsdóttir *et al.* 2009), B. *nana* had already appeared in northern Iceland before the Saksunarvatn tephra fall in 10.2 cal ka BP, whereas B. *pubescens* appeared in the same area around 9.8 cal ka BP with birch (B. *pubescens*-dominated) woodland establishment at approximately 9.3–8.3 (peak 8.7) cal ka BP. Birch arrived in Greenland much later: the dwarf birch B. *nana* was found in eastern Greenland around 7,600 yr BP, whereas the tree-birch species B. *pubescens* only arrived in south-western Greenland sometime before 3,500 yr BP (Fredskild 1991). The Holocene route of birch migration is clearly westwards from Europe, across the Atlantic Ocean to Iceland, and then on to Greenland. In order to elucidate these migration routes better, we plan to study the phylogeography of *Betula* species in the British Isles and Greenland in the next phase of our project.

5. Conclusions

Our botanical, cytogenetic, palynological and molecular studies reviewed here show clearly that hybridisation between B. *nana* and B. *pubescens* is widespread in Iceland; the resulting gene flow via introgressive hybridisation is bi-directional; and that the process is continuous through time and space. Iceland could be considered a birch hybrid zone, harbouring genetic variation which may be advantageous in subarctic regions. Despite extensive introgression across species and ploidy levels, a biogeographical pattern has been observed which indicates different population histories or multiple origins of Icelandic birch. Present-day birch in Iceland is most probably post-glacial in origin, migrating from Western Europe and colonizing Iceland in the early Holocene.

6. Acknowledgments

I am most grateful to Ægir Thór Thórsson for his expertise in the botanical and molecular analysis of Icelandic birch and his dedication to our birch research from early on, and to Lilja Karlsdóttir for her palynological accuracy and dedication. I thank the following people who have contributed to a better understanding of Icelandic birch over the years: Thorsteinn ómasson, Peter Tigerstedt, Áskell and Doris Löve, Pat Heslop-Harrison, Mark Atkinson, ⁄latti Sulkinoja, Adalsteinn Sigurgeirsson, Thorarinn Benediktz, Elina Salmila, Irma aloniemi, Vignir Sigurdsson, Thröstur Eysteinsson, Martin Lascoux, Snæbjörn Pálsson, ⁄largrét Hallsdóttir, Magnús Jóhannsson, Snædís H. Björnsdóttir, Anna Palmé, Kenneth loegh, Ólafur Eggertsson, Bryndís G. Róbertsdóttir, Ploenpit Chokchaichamnankit and 'ridrik R. Jónsson. This birch research has been funded by the Icelandic Research Centre Rannís) and the Research Fund of the University of Iceland.

'. References

\bbott RJ, Smith LC, Milne RI, Crawford RMM, Wolff K, Balfour J (2000) Molecular analysis of plant migration and refugia in the Arctic. *Science* 289: 1343-1346.

\dams RP, Pandey RN, Leverenz JW, Dignard N, Hoegh K, Thorfinnsson T (2003) Pan-Arctic variation in *Juniperus communis*: historical biogeography based on DNA fingerprinting. *Biochemical Systematics and Ecology* 31: 181-192.

\lsos IG, Engelskjon T, Gielly L, Taberlet P, Brochmann C (2005) Impact of ice ages on circumpolar molecular diversity: insights from an ecological key species. *Molecular Ecology* 14: 2739-2753.

\lsos IG, Eidesen PB, Ehrich D, Skrede I, Westergaard K, Jacobsen GH, Landvik JY, Taberlet P, Brochmann C (2007) Frequent long-distance plant colonization in the changing arctic. *Science* 316: 1606-1609.

\namthawat-Jónsson K (1994) Genetic variation in Icelandic birch. *Norwegian Journal of Agricultural Sciences* 18: 9-14.

\namthawat-Jónsson K (2003) Hybrid introgression in Betula. In: Sharma AK, Sharma A, eds. *Plant Genome: Biodiversity and Evolution. Volume 1: Phanerogams*, pp. 249-265. Plymouth, UK; Enfield (NH), USA: Science Publishers, Inc.

\namthawat-Jónsson K (2004) Preparation of chromosomes from plant leaf meristems for karyotype analysis and *in situ* hybridization. *Methods in Cell Science* 25: 91-95.

\namthawat-Jónsson K, Heslop-Harrison JS (1995) Molecular cytogenetics of Icelandic birch species: physical mapping by *in situ* hybridization and rDNA polymorphism. *Canadian Journal of Forest Research* 25: 101-108.

\namthawat-Jónsson K, Thórsson ÆTh (2003) Natural hybridization in birch: Triploid hybrids between *Betula nana* and *B. pubescens*. *Plant Cell Tissue Organ Culture* 75: 99-107.

\namthawat-Jónsson K, Tómasson Th (1990) Cytogenetics of hybrid introgression in Icelandic birch. *Hereditas* 112: 65-70.

\namthawat-Jónsson K, Tómasson T (1999) High frequency of triploid birch hybrid by *Betula nana* seed parent. *Hereditas* 130: 191-193.

\namthawat-Jónsson K, Thórsson ÆTh, Temsch EM, Greilhuber J (2010) Icelandic birch polyploids – the case of perfect fit in genome size. *Journal of Botany* 2010: article ID 347254.

\rnold M. (1992) Natural hybridization as an evolutionary process. *Annual Review of Ecology and Systematics* 23: 237-261.

\tkinson MD (1992) Biological flora of the British Isles, No. 175: *Betula pendula* Roth. (*B. verrucosa* Ehrh.) and *B. pubescens* Ehrh. *Journal of Ecology* 80: 837-870.

Aradóttir ÁL (1991) Population biology and stand development of birch (Betula pubescens Ehrh.) of disturbed sites in Iceland. Ph.D. Dissertation, Texas A&M University, College Station.

Barton NH (2001) The role of hybridization in evolution. *Molecular Ecology* 10: 551-568.

Belahbib N, Pemonge M-H, Ouassou A, Sbay H, Kremer A, Petit RJ (2001) Frequent cytoplasmic exchanges between oak species that are not closely related: *Quercus suber* and *Q. ilex* in Morocco. *Molecular Ecology* 10: 2003-2012.

Belleville B, Cloutier A, Achim A (2011) Detection of red heartwood in paper birch (*Betula payrifera*) using external stem characteristics. *Canadian Journal of Forest Research* 41: 1491-1499.

Bennett KD, Tzedakis PC, Willis KJ (1991) Quaternary refugia of north European trees. *Journal of Biogeography* 18: 103-115.

Bennett MD, Leitch IJ (2011) Nuclear DNA amounts in angiosperms: targets, trends and tomorrow. *Annals of Botany* 107: 467-590.

Bennett MD, Smith JB (1976) Nuclear DNA amounts in angiosperms. *Philosophical Transactions of the Royal Society of London Series B* 274: 227- 274.

Bennike O, Björck S, Böcher J, Hansen L, Heinemeier J, Wohlfarth B (1999) Early Holocene plant and animal remains from North-east Greenland. *Journal of Biogeography* 26: 667-677.

Bersch M, Meincke J, Sy A (1999) Interannual thermohaline changes in the northern North Atlantic 1991-1996. *Deep-Sea Research II* 46: 55-75.

Birks HH, van Dinter M (2010) Lateglacial and early Holocene vegetation and climate gradients in the Nordfjord-Alesund area, western Norway. *Boreas* 39: 783-798.

Bond G, Showers W, Cheseby M, Lotti R, Almasi P, Demenocal P, Priore P, Cullen H, Hajdas I, Bonani G (1997). A pervasive millennial-scale cycle in North Atlantic Holocene and glacial climates. *Science* 14: 1257-1266.

Brochmann C, Gabrielsen TM, Nordal I, Landvik JY, Elven R (2003) Glacial survival or tabula rasa? The history of North Atlantic biota revisited. *Taxon* 52: 417-450.

Brown IR, Kennedy D, William DA (1982) The occurrence of natural hybrids between *Betula pendula* Roth and *B. pubescens* Ehrh. *Watsonia* 14: 133-145.

Brown IR, William DA (1984) Cytology of the *Betula alba* L. complex. Proceedings of the Royal Society of Edinburgh 85B: 49-64.

Buckland PC, Perry DW, Gíslason GM, Dugmore AJ (1981) The pre-Landnám fauna of Iceland: a palaeontological contribution. *Boreas* 15: 173-184.

Caffarra A, Donnelly A, Chuine I (2011) Modelling the timing of *Betula pubescens* budburst II. Integrating complex effects of photoperiod into process-based models. *Climate Research* 46: 159-170.

Caseldine C, Langdon P, Holmes N (2006) Early Holocene climate variability and the timing and extent of the Holocene thermal maximum (HTM) in northern Iceland. *Quaternary Science Review* 25: 2314-2331.

Choler P, Erschbamer B, Tribsch A, Gielly L, Taberlet P (2004) Genetic introgression as a potential to widen a species' niche: Insights from alpine *Carex curvula*. *Proceedings of the National Academy of Sciences of the USA* 101: 171-176.

Clapham AR, Tutin TG, Warburg EF (1962) *Flora of the British Isles, 2nd edition.* Cambridge: Cambridge University Press.

Crautlein M von, Korpelainen H, Pietilainen M, Rikkinen J (2011) DNA barcoding: a tool for improved taxon identification and detection of species diversity. *Biodiversity Conservation* 20: 373-389.

Demesure B, Sodzi N, Petit RJ (1995) A set of universal primers for amplification of polymorphic noncoding regions of mitochondrial and chloroplast DNA in plants. *Molecular Ecology* 4: 129-131.

Deslippe JR, Hartmann M, Mohn WW, Simard SW (2011) Long-term experimental manipulation of climate alters the ectomycorrhizal community of *Betula nana* in Arctic tundra. *Global Change Biology* 17: 1625-1636.

Drouin M, Beauregard R, Duchesne I (2010) Impact of paper birch (*Betula papyrifera*) tree characteristics on lumber color, grade recovery, and lumber value. *Forest Products Journal* 60: 236-243.

de Groot WJ, Thomas PA, Wein RW (1997) *Betula nana* L. and *Betula glandulosa* Michx. *Journal of Ecology* 85: 241-264.

Ejankowski W (2010) Demographic variation of dwarf birch (*Betula nana*) in communities dominated by *Ledum palustre*. *Biologia* 65: 248-253.

Elkington TT (1968) Introgressive hybridisation between *Betula nana* L. and *B. pubescens* Ehrh. in northwest Iceland. *New Phytologist* 67: 109-118.

Ellstrand NC, Prentice HC, Hancock JF (1999) Gene flow and introgression from domestic plants into their wild species. *Annual Review of Ecology and Systematics* 30: 539-563.

Enkhtuya B, Oskarsson U, Dodd JC, Vosadka M (2003) Inoculation of grass and tree seedlings used for reclaiming eroded areas in Iceland with mycorrhizal fungi. *Folia Geobotanica* 38: 209-222.

Eriksson G, Jonsson A (1986) A review of the genetics of *Betula*. *Scandinavian Journal of Forest Research* 1: 421-434.

Erlendsson E, Edwards KJ (2009) The timing and causes of the final pre-settlement expansion of *Betula pubescens* in Iceland. *Holocene* 19: 1083-1091.

Erler A, Hawranek T, Kruckemeier L, Asam C, Egger M, Ferreira F, Briza P (2011) Proteomic profiling of birch (*Betula verrucosa*) pollen extracts from different origins. *Proteomics* 11: 1486-1498.

Franiel I, Babczynska A (2011) The growth and reproductive effort of *Betula pendula* Roth in heavy-metals polluted area. *Polish Journal of Environmental Studies* 20: 1097-1101.

Fredskild B (1991) The genus *Betula* in Greenland – Holocene history, present distribution and synecology. *Nordic Journal of Botany* 11: 393-412.

Freysdottir J, Sigurpalsson MB, Omarsdottir S, Olafsdottir ES, Vikingsson A, Hardardottir I (2011) Ethanol extract from birch bark (*Betula pubescens*) suppresses human dendritic cell mediated Th1 responses and directs it towards a Th17 regulatory response *in vitro*. *Immunology letters* 136: 90-96.

Furlow JJ (1997) Betulaceae Gray. Birch family. In: Flora of North America Committee, eds. *Flora of North America, north of Mexico, vol. 3*, pp. 507-538. New York: Oxford University Press.

Fussi B, Lexer C, Heinze B (2010) Phylogeography of *Populus alba* (L) and *Populus tremula* (L) in Central Europe: secondary contact and hybridisation during reolonisation from disconnected refugia. *Tree Genetics & Genomics* 6: 439-450.

Gardiner AS (1984) Taxonomy of infraspecific variation in *Betula pubescens* Ehrh., with particular reference to the Scottish Highlands. *Proceedings of the Royal Society of Edinburgh* 85B: 13-26.

Gardner S, Sidisunthorn P, Anusarnsunthorn V (2000) *A Field Guide to Forest Trees of Northern Thailand*. Bangkok: Kobfai Publishing Project. ISBN 974-7798-29-8.

Geirsdóttir Á, Hardardóttir J, Sveinbjörndóttir ÁE (2000) Glacial extent and catastrophic meltwater events during the deglaciation of Southern Iceland. *Quaternary Science Reviews* 19: 1749-1761.

Gröntved J (1942) The Pteridophyta and Spermatophyta of Iceland: Betulaceae. *The Botany of Iceland* 4: 206-209.

Grundermann C, Gruber CW, Hertrampf A, Zehl M, Kopp B, Huber R (2011) An aqueous birch leaf extract of *Betula pendula* inhibits the growth and cell division of inflammatory lymphocytes. *Journal of Ethnopharmacology* 136: 444-451.

Guo JJ, Zeng J, Zhou SL, Zhao ZG (2008) Isolation and characterization of 19 microsatellite markers in a tropical and warm subtropical birch, *Betula alnoides* Buch.-Ham. ex. D. Don. *Molecular Ecology Resources* 8: 895-897.

Hannon GE, Rundgren M, Jessen CA (2010) Dynamic early Holocene vegetation development on the Faroe Islands inferred from high-resolution plant macrofossil and pollen data. *Quaternary Research* 73: 163-172.

Hallsdóttir M (1995) On the pre-settlement history of Icelandic vegetation. *Icelandic Agricultural Sciences* 9: 17-29.

Hallsdóttir M, Caseldine CJ (2005) The Holocene vegetation history of Iceland, state-of-the-art and future research. In: Caseldine C, Russell A, Hardardóttir J, Knudsen Ó, eds. *Iceland – modern processes and past environments*, pp. 319-334. Amsterdam: Elsevier.

Heiser CB Jr (1973) Introgression re-examined. *Botanical Review* 39: 347-366.

Huh JE, Hong JM, Baek YH, Lee JD, Choi DY, park DS (2011) Anti-inflammatory and anti-nociceptive effect of *Betula platyphylla* var. *japonica* in human interleukin-1 beta-stimulated fibroblast-like synoviocytes and in experimental animal models. *Journal of Ethnopharmacology* 135: 126-134.

Hultén E, Fries M (1986) *Atlas of North European Vascular Plants*. Köningstein: Koeltz Scientific Books.

Hynynen J, Niemisto P, Vihera-Aarnio A, Brunner A, Hein S, Velling P (2010) Silviculture of birch (*Betula pendula* Roth and *Betula pubescens* Ehrh.) in northern Europe. *Forestry* 83: 103-119.

Jarvis DI, Hodgkin T (1999) Wild relatives and crop cultivars: detecting natural introgression and farmer selection of new genetic combinations in agroecosystems. *Molecular Ecology* 8: S159-S173.

Jetlund S (1994) Introgressive hybridisation between the birch species (*Betula pubescens ssp. tortuosa*) and *Betula nana* in the mountains in "Gudbrandsdalen", Norway. *Norwegian Journal of Agricultural Sciences* 18: 15-18.

Jiang TB, Zhou BR, Gao FL, Gao BZ (2011) Genetic linkage maps of white birches (*Betula platyphylla* Suk. and *B. pendula* Roth) based on RAPD and AFLP markers. *Molecular Breeding* 27: 347-356.

Johansen S, Hytteborn H (2001) A contribution to the discussion of biota dispersal with drift ice and driftwood in the North Atlantic. *Journal of Biogeography* 28: 105-115.

Johnsson H (1945) Interspecific hybridisation within the genus *Betula*. *Hereditas* 31: 163-176.

Kallio P, Niemi S, Sulkinoja M (1983) The Fennoscandian birch and its evolution in the marginal forest zone. *Nordicana* 47: 101-110.

Karlsdóttir L, Hallsdóttir M, Thórsson ÆTh, Anamthawat-Jónsson K (2008) Characteristics of pollen from natural triploid *Betula* hybrids. *Grana* 47: 52-59.

Karlsdóttir L, Hallsdóttir M, Thórsson ÆTh, Anamthawat-Jónsson K (2009) Evidence of hybridisation between *Betula pubescens* and *B. nana* in Iceland during the early Holocene. *Rev Palaeobotany Palynology* 156: 350-357.

Karlsdóttir L, Thórsson ÆTh, Hallsdóttir M, Sigurgeirsson A, Eysteinsson Th, Anamthawat-Jónsson K (2007) Differentiating pollen of *Betula* species from Iceland. *Grana* 46: 78-84.

Kaufman DS, Ager TA, Anderson NJ, Anderson PM, Andrews JT, Bartlein PJ, Brubaker LB, Coats LL, Cwynar LC, Duvall ML, Dyke AS, Edwards ME, Eisner WR, Gajewski K, Geirsdóttir A, Hu FS, Jennings AE, Kaplan MR, Kerwin MW, Lozhkin AV, MacDonald GM, Miller GH, Mock CJ, Oswald WW, Otto-Bliesner BL, Porinchu DF, Ruüland K, Smol JP, Steig E J, Wolfe BB (2004) Holocene thermal maximum in the western Arctic (0-180°W). *Quaternary Science Reviews* 23: 529-560.

Kenworthy JB, Aston D, Bucknall SA (1972) A study of hybrids between *Betula pubescens* Ehrh. and *Betula nana* L. from Sutherland – an integrated approach. *Transactions of the Botanical Society of Edinburgh* 42: 517-539.

Kinnaird JW (1974) Effects of site conditions on the regeneration of birch (*Betula pendula* Roth and *B. pubescens* Ehrh.). *Journal of Ecology* 62: 467-473.

Kleczewski NM, Herms DA, Bonello P (2010) Effects of soil type, fertilization and drought on carbon allocation to root growth and partitioning between secondary metabolism and ectomycorrhizae of *Betula papyrifera*. *Tree Physiology* 30: 807-817.

Kontunen-Soppela A, Parvianen J, Ruhanen H, Brosche M, Keinanen M, Thakur RC, Kolehmainen M, Kangasjarvi J, Oksanen E, Karnosky DF, Vapaavuori E (2009) Gene expression responses of paper birch (*Betula papyrifera*) to elevated CO_2 and O_3 during leaf maturation and senescence. *Environmental Pollution* 158: 959-968.

Kristinsson H (1995) Post-settlement history of Icelandic forests. *Icelandic Agricultural Sciences* 9: 31-35.

Kund M, Vares A, Sims A, Tullus H, Uri V (2010) Early growth and development of silver birch (*Betula pendula* Roth.) plantations on abandoned agricultural land. *European Journal of Forest Research* 129: 679-688.

Kuoppamaa M, Huusko A, Hicks S (2009) *Pinus* and *Betula* pollen accumulation rates from the northern boreal forest as a record of interannual variation in July temperature. *Journal of Quaternary Science* 24: 513-521.

Levanic T, Eggertsson O (2008) Climatic effects on birch (*Betula pubescens* Ehrh.) growth in Fnjoskadalur valley, northern Iceland. *Dendrochronology* 25: 135-143.

Li JH, Shoup S, Chen ZD (2007) Phylogenetic relationships of diploid species of *Betula* (Betulaceae) inferred from DNA sequences of nuclear nitrate reductase. *Systematic Botany* 32: 357-365.

Li PQ, Cheng SX (1979) Betulaceae. In: Kuang KZ, Li PQ, eds. *Flora Republicae Popularis Sinicae*, vol. 21, pp. 44-137. Beijing: Science Press.

Li PQ, Skvortsov AK (1999) Betulaceae. In: Wu ZY, Raven PH, eds. *Flora of China, vol. 4: Cycadaceae through Fagaceae*, pp. 286–313. Beijing: Science Press; St. Louis: Missouri Botanical Garden Press.

Linkosalo T, Ranta H, Oksanen A, Siljamo P, Luomajoki A, Kukkonen J, Sofiev M (2010) A double-threshold temperature sum model for predicting the flowering duration and relative intensity of *Betula pendula* and *B. pubescens*. *Agricultural and Forest Meteorology* 150: 1579-1584.

Lintunen A, Sievanen R, Kaitaniemi P, Perttunen J (2011) Models of 3D crown structure for Scots pine (*Pinus sylvestris*) and silver birch (*Betula pendula*) growth in mixed forest. *Canadian Journal of Forest Research* 41: 1779-1794.

Löve Á, Löve D (1956) A cytotaxonomical conspectus of the Icelandic flora. *Acta Horti Gothobergensis* 20: 65-290.

Luostarinen K, Huotari N, Tillman-Sutela E (2009) Effect of regeneration method on growth, wood density and fibre properties of downy birch (*Betula pubescens* Ehrh.) *Silva Fennica* 43: 329-328.

Mandre M, Parn H, Kloseiko J, Ingerslev M, Stupak I, Kort M, Paasrand K (2010) Use of biofuel ashes for fertilisation of *Betula pendula* seedlings on nutrient-poor peat soil. *Biomass & Bioenergy* 34: 1384-1392.

Maliouchenko O, Palmé AE, Buonamici A, Vendramin GG, Lascoux M (2007) Comparative phylogeography and population structure of European *Betula* species, with particular focus on *B. pendula* and *B. pubescens*. *Journal of Biogeography* 34: 1601-1610.

Migalina SV, Ivanova LA, Makhnev AK (2010) Changes of leaf morphology in *Betula pendula* Roth and *B. pubescens* Ehrh. along a zonal-climatic transect in the Urals and Western Siberia. *Russian Journal of Ecology* 41: 293-301.

Minder AM, Widmer A (2008) A population genomic analysis of species boundaries: neutral processes, adaptive divergence and introgression between two hybridizing plant species. *Molecular Ecology* 17: 1552-1563.

Morales LO, Tegelberg R, Brosche M, Lindfors A, Siipola S, Aphalo PJ (2011) Temporal variation in epidermal flavanoids due to altered solar UV radiation is moderated by the leaf position in *Betula pendula*. *Physiologia Plantarum* 143: 261-270.

Nagamitsu T, Kawahara T, Kanazashi A (2006) Endemic dwarf birch *Betula apoiensis* (Betulaceae) is a hybrid that originated from *Betula ermanii* and *Betula ovalifolia*. *Plant Species Biology* 21: 19-29.

Nielsen DG, Muilenburg VL, Herms DA (2011) Interspecific variation in resistance of Asian, European, and North American birches (*Betula* spp.) to bronze birch borer (Coleoptera: Buprestidae). *Environmental Entomology* 40: 648-653.

Norddahl H (1991) A review of the glaciation maximum concept and the deglaciation of Eyjafjördur, North Iceland. In: Maizels JK, Caseldine C, eds. *Environmental change in Iceland: past and present*, pp. 31-47. Dordrecht: Kluwer Academic Publishers.

Norddahl H, Pétursson HG (2005) Relative sea-level changes in Iceland; new aspects of the Weichselian deglaciation of Iceland. In: Caseldine C, Russel A, Hardardottir J, Knudsen O, eds. *Iceland – Modern processes and past environments*, pp. 25-78. Amsterdam: Elsevier.

Oddsdottir ES, Eilenberg J, Sen R, Harding S, Halldorsson G (2010) Early reduction of *Otiorhynchus* spp. larval root herbivory on *Betula pubescens* by beneficial soil fungi. *Applied Soil Ecology* 45: 168-175.

Palmé AE, Su Q, Palsson S, Lascoux M (2004) Extensive sharing of chloroplast haplotypes among European birches indicates hybridization among *Betula pendula*, *B. pubescens* and *B. nana*. *Molecular Ecology* 13: 167-178.

Paus A (1995) The Late Weichselian and early Holocene history of tree birch in south Norway and the Bolling *Betula* time-lag in northwest Europe. *Review of Palaeobotany and Palynology* 85: 243-262.

Petit RJ, Csaikl UM, Bordacs S, Burg K, Coart E, Cottrell J, van Dam B, Deans JD, Dumolin-Lapegue S, Fineschi S, Finkeldey R, Gillies A, Glaz I, Goicoechea PG, Jensen JS, Konig AO, Lowe AJ, Madsen SF, Matyas SF, Munro RC, Olalde M, Pemonge MH, Popescu F, Slade D, Tabbener H, Taurchini D, de Vries SGM, Ziegenhagen B, Kremer A (2002) Chloroplast DNA variation in European white oaks: Phylogeography and patterns of diversity based on data from over 2600 populations. *Forest Ecology and Management* 156: 5-26.

Phan MG, Thi TCT, Phan TS, Matsunami K, Otsuka H (2011) Three new dammarane glycocides from *Betula alnoides*. *Phytochemistry Letters* 4: 179-182.

Pudas E, Leppala M, Tolvanen A, Poikolainen J, Venalainen A, Kubin E (2008) Trends in phenology of *Betula pubescens* across the boreal zone in Finland. International *Journal of Biometeorology* 52: 251-259.

Quinn GP, Keough MJ (2002) *Experimental Design and Data Analysis for Biologists*. Cambridge: Cambridge University Press.

Rea RV (2011) Impacts of moose (*Alces alces*) browsing on paper birch (*Betula papyrifera*) morphology and potential timber quality. *Silva Fennica* 45: 227-236.

Reece JB, Urry LA, Cain ML, Wasserman SA, Minorsky PV, Jackson RB (2011) *Campbell Biology*, 9th edition. Pearson USA.

Rieseberg LH (1997) Hybrid origins of plant species. *Annual Review of Ecology and Systematics* 28: 359-389.

Rieseberg LH, Willis JH (2007) Plant speciation. *Science* 317: 910-914.

Rundgren M (1998) Early-Holocene vegetation of northern Iceland: pollen and plant macrofossil evidence from the Skagi peninsula. *The Holocene* 8: 553-564.

Ruuhola T, Leppanen T, Lehto T (2011) Retranslocation of nutrients in relation to boron availability during leaf senescence of *Betula pendula* Roth. *Plant and Soil* 344: 227-240.

Sanz R, Pulido F, Camarero JJ (2011) Boreal trees in the Mediterranean: recruitment of downy birch (*Betula alba*) at its southern limit. *Annals of Forest Science* 68: 793-802.

Sano M, Furuta F, Sweda T (2010) Summer temperature variations in southern Kamchatka as reconstructed from a 247-year tree ring chronology of *Betula ermanii*. *Journal of Forest Research* 15: 234-240.

Schenk MF, Thienpont C-N, Koopman WJM, Gilissen LJWJ, Smulders MJM (2008) Phylogenetic relationships in *Betula* (Betulaceae) based on AFLP markers. *Tree Genetics & Genomics* 4: 911-924.

Schofield JE, Edwards KJ (2011) Grazing impacts and woodland management in *Eriksfjord* *Betula*, coprophilous fungi and the Norse settlement in Greenland. *Vegetation History and Archaeobotany* 20: 181-197.

Saeki I, Dick CW, Barnes BV, Murakami N (2011) Comparative phylogeography of red maple (*Acer rubrum* L.) and silver maple (*Acer saccharinum* L.): impacts of habitat specialization, hybridization and glacial history. *Journal of Biogeography* 38: 992-1005.

Shrestha BB, Ghimire B, Lekhak HD, Jha PK (2007) Regeneration of treeline birch (*Betula utilis* D. Don) forest in a Trans-Himalayan dry valley in central Nepal. *Mountain Research and Development* 27: 259-267.

Skrede I, Eidesen PB, Portela RP, Brochmann C (2006) Refugia, differentiation and postglacial migration in arctic-alpine Eurasia, exemplified by mountain avens (*Dryas octopetala* L.). *Molecular Ecology* 15: 1827-1840.

Sokal RR, Rohlf FJ (1995) *Biometry, 3rd edition*. New York: WH Freeman.

Stefánsson S (1901) 18. Betuláceæ (Bjarkættin). In: Möller SL, ed. *Flóra Íslands*, pp. 73-75 Copenhagen: Hin Íslenzka Bókmenntafjelag.

Sulkinoja M (1990) Hybridization, introgression and taxonomy of the mountain birch in SW Greenland compared with related results from Iceland and Finnish Lapland *Meddelelser om Grønland Bioscience* 33: 21-29.

Tabata A, Ono K, Sumida A, Hara T (2010) Effects of soil water conditions on the morphology, phenology, and photosynthesis of *Betula ermanii* in the boreal forest. *Ecological Research* 25: 823-835.

Taberlet P, Fumagalli L, Wust-Saucy A-G, Cosson J-F (1998) Comparative phylogeography and postglacial colonization routes in Europe. *Molecular Ecology* 7: 453-464.

Taberlet P, Gielly L, Pautou G, Bouvet J (1991) Universal primers for amplification of three noncoding regions of chloroplast DNA. *Plant Molecular Biology* 17: 1105-1109.

Takahashi K, Uemura S, Hara T (2011) A forest-structure-based analysis of rain flow into soil in a dense deciduous *Betula ermanii* forest with understory dwarf bamboo. *Landscape and Ecological Engineering* 7: 101-108.

Taper LJ, Grant WF (1973) The relationship between chromosome size and DNA content in birch (*Betula*) species. *Caryologia* 26: 263-273.

Tarasov PE, Volkova VS, Webb T III, Guiot J, Andreev AA, Bezusko LG, Bezusko TV, Bykova GV, Dorofeyuk NI, Kvavadze EV, Osipova IM, Panova NK, Sevastyanov DV (2000) Last glacial maximum biomes reconstructed from pollen and plant macrofossil data from northern Eurasia. *Journal of Biogeography* 27: 609-620.

Thórsson ÆTh (2008) Genecology, introgressive hybridisation and phylogeography of *Betula* species in Iceland. Ph.D. Dissertation, University of Iceland. ISBN 978-9979-70-481-2.

Thórsson Æ, Salmela E, Anamthawat-Jónsson K (2001) Morphological, cytogenetic, and molecular evidence for introgressive hybridization in birch. *Journal of Heredity* 92: 404-408.

Thórsson ÆTh, Pálsson S, Sigurgeirsson A, Anamthawat-Jónsson K (2007) Morphological variation among *Betula nana* (diploid), *B. pubescens* (tetraploid) and their triploid hybrids in Iceland. *Annals Botany* 99: 1183-1193.

Thórsson ÆTh, Pálsson S, Lascoux M, Anamthawat-Jónsson K (2010) Introgression and phylogeography of *Betula nana* (diploid), *B. pubescens* (tetraploid) and their triploid hybrids in Iceland inferred from cp-DNA haplotype variation. *Journal of Biogeography* 37: 2098-2110.

Truong C, Palmé AE, Felber F (2007) Recent invasion of the mountain birch *Betula pubescens* ssp. *tortuosa* above the treeline due to climate change: genetic and ecological study in northern Sweden. *Journal of Evolutionary Biology* 20: 369-380.

Tsuda Y, Ide Y (2005) Wide-range analysis of genetic structure of *Betula maximowicziana*, a long-lived pioneer tree species and noble hardwood in the cool temperate zone of Japan. *Molecular Ecology* 14: 3929-3941.

Tsuda Y, Ide Y (2010) Chloroplast DNA phylogeography of *Beutla maximowicziana*, a long-lived pioneering tree species and noble hardwood in Japan. *Journal of Plant Research* 123: 343-353.

Vaarama A, Valanne T (1973) On the taxonomy, biology and origin of *Betula tortuosa* Ledeb. *Reports of the Kevo Subarctic Research Station* 10: 70-84.

Vasari Y (1972) The history of the vegetation in Iceland during the Holocene. In: Vasari Y, Hyvärinen H, Hicks S, eds. *Climatic changes in Arctic areas during the last ten-thousand years*, pp. 239-252. Finland: University of Oulu.

Walters SM (1964) Betulaceae. In: Tutin TG, Heywood VH, Burges NA, Valentine DH, Walters SM, Webb DA, eds. *Flora Europaea, vol. 1*, pp. 57-59. Cambridge: Cambridge University Press.

Wastl M, Stötter J, Caseldine C (2001) Reconstruction of Holocene variations of the upper limit of tree or shrub birch growth in northern Iceland based on evidence from Vesturárdalur-Skídadalur, Tröllaskagi. *Arctic, Antarctic and Alpine Research* 33: 191-203.

Westergaard KB, Alsos IG, Popp M, Engelskjon T, Flatberg KI, Brochmann C (2011a) Glacial survival may matter after all: nunatak signature in the rare European populations of two west-arctic species. *Molecular Ecology* 20: 376-393.

Westergaard KB, Alsos IG, Engelskjon T, Flatberg KI, Brochmann C (2011b) Trans-Atlantic genetic uniformity in the rare snowbed sedge *Carex rufina*. *Conservation Genetics* 12: 1367-1371.

Willis KJ, Rudner E, Sumegi P (2000) The full-glacial forests of central and south-eastern Europe. *Quaternary Research* 53: 203-213.

Woodworth RH (1929) Cytological studies in the Betulaceae. I. *Betula*. *The Botanical Gazette* 87: 331-363.

Xiong JA, Taniguchi M, Kashiwada Y, Yamagishi T, Takaishi Y (2011) Seven new dammarane triterpenes from the floral spikes of *Betula platyphilla* var. *japonica*. *Journal of Natural Medicines* 65: 217-223.

Zeng J, Li J-H, Chen Z-D (2008) A new species of *Betula* section *Betulaster* (Betulaceae) from China. *Botanical Journal of the Linnean Society* 156: 523-528.

Zobel DB, Singh SP (1997) Himalayan forests and ecological generalizations. *BioScience* 47: 735-745.

Permissions

The contributors of this book come from diverse backgrounds, making this book a truly international effort. This book will bring forth new frontiers with its revolutionizing research information and detailed analysis of the nascent developments around the world.

We would like to thank Kesara Anamthawat-Jónsson, for lending her expertise to make the book truly unique. She has played a crucial role in the development of this book. Without her invaluable contribution this book wouldn't have been possible. She has made vital efforts to compile up to date information on the varied aspects of this subject to make this book a valuable addition to the collection of many professionals and students.

This book was conceptualized with the vision of imparting up-to-date information and advanced data in this field. To ensure the same, a matchless editorial board was set up. Every individual on the board went through rigorous rounds of assessment to prove their worth. After which they invested a large part of their time researching and compiling the most relevant data for our readers. Conferences and sessions were held from time to time between the editorial board and the contributing authors to present the data in the most comprehensible form. The editorial team has worked tirelessly to provide valuable and valid information to help people across the globe.

Every chapter published in this book has been scrutinized by our experts. Their significance has been extensively debated. The topics covered herein carry significant findings which will fuel the growth of the discipline. They may even be implemented as practical applications or may be referred to as a beginning point for another development. Chapters in this book were first published by InTech; hereby published with permission under the Creative Commons Attribution License or equivalent.

The editorial board has been involved in producing this book since its inception. They have spent rigorous hours researching and exploring the diverse topics which have resulted in the successful publishing of this book. They have passed on their knowledge of decades through this book. To expedite this challenging task, the publisher supported the team at every step. A small team of assistant editors was also appointed to further simplify the editing procedure and attain best results for the readers.

Our editorial team has been hand-picked from every corner of the world. Their multi-ethnicity adds dynamic inputs to the discussions which result in innovative outcomes. These outcomes are then further discussed with the researchers and contributors who give their valuable feedback and opinion regarding the same. The feedback is then collaborated with the researches and they are edited in a comprehensive manner to aid the understanding of the subject.

Apart from the editorial board, the designing team has also invested a significant amount of their time in understanding the subject and creating the most relevant covers. They scrutinized every image to scout for the most suitable representation of the subject and create an appropriate cover for the book.

The publishing team has been involved in this book since its early stages. They were actively engaged in every process, be it collecting the data, connecting with the contributors or procuring relevant information. The team has been an ardent support to the editorial, designing and production team. Their endless efforts to recruit the best for this project, has resulted in the accomplishment of this book. They are a veteran in the field of academics and their pool of knowledge is as vast as their experience in printing. Their expertise and guidance has proved useful at every step. Their uncompromising quality standards have made this book an exceptional effort. Their encouragement from time to time has been an inspiration for everyone.

The publisher and the editorial board hope that this book will prove to be a valuable piece of knowledge for researchers, students, practitioners and scholars across the globe.

List of Contributors

Graciela García
Evolutionary Genetics Section, Biology Institute, Faculty of Sciences, UdelaR, Montevideo, Uruguay

M. Ruiz-García, C. Vásquez and M. Pinedo-Castro
Molecular Genetics Population- Evolutionary Biology Laboratory, Genetics Unit, Biology Department, Science Faculty, Pontificia Javeriana University, Bogota DC, Colombia

S. Sandoval
Tapir Preservation Fund, Bogotá DC, Colombia

A. Castellanos
Fundación Espíritu del Bosque, c/ Barcelona 311 y Tolosa, Quito, Equador

F. Kaston
Fundación Nativa, Colombia

B. de Thoisy
Association Kwata, BP 672, 97335 Cayenne cedex, French Guiana

J. Shostell
Biology Department, Penn State University-Fayette, Uniontown, Pennsylvania, USA

Samuel K. Sheppard
The University of Oxford, Department of Zoology, United Kingdom
The University of Swansea, College of Medicine, United Kingdom

Helen M. L. Wimalarathna
The University of Oxford, Department of Zoology, United Kingdom

Ingi Agnarsson
University of Puerto Rico, Puerto Rico, USA
Department of Entomology, National Museum of Natural History, Smithsonian Institution, USA

Matjaž Kuntner
Department of Entomology, National Museum of Natural History, Smithsonian Institution, USA
Institute of Biology, Scientific Research Centre, Slovenian Academy of Sciences and Arts, Slovenia

Kesara Anamthawat-Jónsson
Institute of Life and Environmental Sciences, University of Iceland, Askja, Reykjavik, Iceland

Printed in the USA
CPSIA information can be obtained
at www.ICGtesting.com
JSHW011333221024
72173JS00003B/147

9 781632 395115